Flushed with Pride

Also by Wallace Reyburn

Novels
Follow a Shadow
Port of Call
The Street that Died
Three Women
Good and Evil
Getting the Boy

War
Rehearsal for Invasion
Some of It was Fun

Sport
The World of Rugby
The Lions
The Unsmiling Giants
Best Rugby Stories (editor)
The Rugby Companion

Biography
Frost – Anatomy of a Success

FLUSHED WITH PRIDE

The Story of
Thomas Crapper

Wallace Reyburn

PRENTICE-HALL, INC.
Englewood Cliffs, N.J.

Flushed with Pride: The Story of Thomas Crapper
by Wallace Reyburn
First published in the U.S.A. by Prentice-Hall, Inc.,
    Englewood Cliffs, 1971
First published in Great Britain by Macdonald and Company
    (Publishers) Ltd., 1969
© 1969 by Wallace Reyburn
Library of Congress Catalog Card Number: 73-129143
Printed in the United States of America • *T*
ISBN-0-13-322560-7
Prentice-Hall International, Inc., London
Prentice-Hall of Australia, Pty. Ltd., Sydney
Prentice-Hall of Canada, Ltd., Toronto
Prentice-Hall of India Private Ltd., New Delhi
Prentice-Hall of Japan, Inc., Tokyo

Contents

Flushed with Pride

Chapter 1
Unsung Hero

Never has the saying 'A prophet is without honour in his own land' been more true than in the case of Thomas Crapper. Here was a man whose foresight, ingenuity and perseverance brought to perfection one of the great boons to mankind. But is his name revered in the same way as, for example, that of the Earl of Sandwich?

Whenever we order a sandwich we pay verbal tribute to the Earl who originated the idea, to enable him to spend more time at the gaming tables. *He* had been accorded the ultimate tribute to man's inventiveness – his name incorporated into the language without a capital letter.

Likewise with Lord Chesterfield, the Earl of Davenport, Lord Cardigan, the Duke of Wellington – *chesterfield, davenport, cardigan* and *wellingtons* are all everyday words now. Nobody ever had any hesitation in saying, 'Would you like to sit on the chesterfield?' But what of poor Crapper? 'Would you like to wash your hands?' is a pretty poor tribute to a man who put in as much effort, if not more, into *his* contribution to human comfort.

It might be argued that snobbery was the reason why *crapper* was not to become an accepted word. Sandwich and the others were all of the nobility, whereas Thomas Crapper was a mere commoner. But that argument does not stand up when we consider *macintoshes, macadam, bowlers, gladstones, watts, bloomer*s and all the other verbal tributes to commoners such as he, although in passing it should be noted that perhaps there is discrimination against women, since Mrs Bloomer is the only woman thus honoured. Thomas Crapper's fellow-countrymen have also hastened to give full credit to inventors of other lands with such words as *diesel, fahrenheit, hoover, ohms, volts* and *stetson*. We cannot get through a day without giving some such recognition to a pioneer, British or foreign. But not poor Crapper.

It was left to the Americans to give the man his due, and ironically enough these American word coinages are now finding their way back across the Atlantic, for the British to pay indirect, and belated tribute to the Chelsea pioneer.

Thomas Crapper was born of a humble family in the Yorkshire town of Thorne in 1837 – appropriately enough the year in which Queen Victoria came to the throne, the start of the Victorian Era to which he was to make such a splendid contribution.

His grand-niece Edith Crapper, who is the last direct link with this family and who was to be of great help in my researches, mentioned in a letter to me that she understood that their forbears had come from Holland and in their time the name had been spelt with a 'K'. 'But,' she added, 'I think spelling it Krapper makes the name look ugly.'

The *Dictionary of British Surnames*, however, has it as a very old English name, well known in Yorkshire for centuries. The original version was Cropper and – as with Butcher, Baker, Carpenter and so on – it was a name of trade or calling, someone who brings in the crops. According to the dictionary the first recorded variation with an 'a' was in a parish register of 1315, which contained the name of Alice Le Crappere.

Nevertheless grand-niece Edith might well be right about the Dutch derivation. Thorne, near Doncaster, stands on marsh land which was even more moist in the old days, when 'fowlers and fishermen lived on dry patches of the fens and marshes.' In 1609 Cornelius Vermuyden, a famous Dutch engineer of the period, was called in to drain the area and he brought over Dutch Protestants to handle the work.

Whatever their ancestry, the early part of last century saw the Crappers a solid, if none too prosperous, Yorkshire family. Tom's father, Chas. was a seafaring man who did not bring in a great deal of money. In 1825 Ma Sarah had to go in arrears with the rates to the tune of 1s. 10½d. And there was hardly a year on through to the 1830s that they weren't behind, owing as much as £1 0s. 2d. in the April assessment of 1833. However, she somehow managed to raise five sons, who worked in and around the shipbuilding yards of Thorne Quay on the River Don. As today, this was, with mining, the main source of employment for the men of Thorne and it seems that around the middle of the 19th century jobs became hard to come by.

Young Thomas Crapper soon felt it necessary to look elsewhere for regular work.

When I started delving into the life story of our man I unearthed two pieces of information which, when taken side by side, just didn't seem to make sense:

(a) he was born in 1837;

(b) he walked from Yorkshire to London in search of work in 1848.

By simple arithmetic this meant that he had walked the 165 miles from Thorne and gone to work as a plumber at the age of 11. Obviously, I thought, one of the dates was wrong and I spent some time checking and double checking them. But any expert of life in Early Victorian times could have told me I was expending a lot of unnecessary energy. Although difficult to comprehend today, the working age of 11 was not out of the ordinary for that period.

To quote from J. H. Clapham in *Early Victorian*

3

E.B.CRAPPER

*Thomas Crapper on his way from Yorkshire to
London in 1848*   Drawing by Edith Crapper

England 1830–1865, 'the Parliamentary report of 1843 ascertained that, taking the country as a whole, regular work generally began between the ages of seven and eight; so that the attempt of the earlier Factory Act to standardise the starting age at nine was not merely the rectification of an abuse, but a reform in average industrial practice.'

However, that long walk to London at the tender age of 11 could not have been something that was done by many boys and reflects great courage on the part of young Thomas. We would not wish that sort of thing upon youngsters today, but it is interesting to note that while Thomas Crapper was presumably 'a victim of the vicious child labour system' it appeared to do him no lasting harm. He spent an energetic, worthwhile life, became prosperous, lived to a ripe 73 years of age and in his time if he didn't walk with Kings he at least discussed sanitary arrangements with them.

Chapter 2
The Chelsea Plumber

When young Crapper came to London in 1848 he got a job with a master plumber in Robert Street in Chelsea, the Royal Borough to which he was to remain loyal throughout his life. Robert Street was off the King's Road, directly opposite the Chelsea Town Hall.

In passing it should be mentioned that in that year of 1848, England's Great Man of cricket, W. G. Grace, was born. Which would appear to have no connection with Thomas Crapper. But, as will be seen later, the interesting thing is that they were to wind up close together.

Young Tom lived in the attic at Robert Street. In those days you lived where you worked or within walking distance. Considering that he was getting only 4s. a week, the penny omnibus was not for the working man, let alone the apprentice. Later in life he told a friend that a vivid memory from those young days was chapped hands and chilblains from constantly messing about with water and from the numbing cold when he would try to get some life back into his hands by holding them over the open grate. On the coldest winter nights he was allowed to take up to bed with him a hot brick from the oven wrapped in flannel.

He remembered being much impressed by the coming of electric light, which, if one looks back over their issues of 1848, was not the case with the *Illustrated London News*, 'The light produced was of the most powerful character, but it is, in our opinion, still but a costly experimental toy, whose practicability forms a whole subject for conjecture,' commented that worthy journal.

Those old copies of the *Illustrated London News* reflect, more accurately, a turbulent world in that year of 1848 when the youthful Crapper came to the big city. Revolution in France, rebellions in Italy, Switzerland, West Germany, Austria and Poland, not to mention the Americans battling it out with the Mexicans ('How is the war to be carried on? How is it to be paid for?').

The Chartists, the working class people who wanted their voice to be heard, were the main trouble-makers in England. At their 'monster demonstration' in Hyde Park on April 15, 1848, 'a double file of Metropolitan police and Pensioners, under arms and fully accoutred,' were on hand to try to keep them under control in the area of Apsley House, 'where the bullet proof shutters were up'. With the Chartist delegates shouting, 'We have held a meeting that had been forbidden and held down!' there were then 'symptoms of unruliness among the crowd that showed itself in violent rushes made from one point to another,' and 'the pressure of the concourse was so great that the lines of police were forced and many of them were carried along with the throng.'

There is no way of knowing whether young Tommy Crapper was present at this or any of the other demonstrations of more than a hundred years ago which had a strangely 1960s flavour about them. He would most certainly have been working the 64-hour week that the Chartists were complaining about but since there is no evidence of his having a militant streak one can only conclude he just got on with it, determined to make a go of it as a plumber.

The Chelsea in which he was soon to be working on his own had an abundance of the charm which, in much modified form, has still managed to

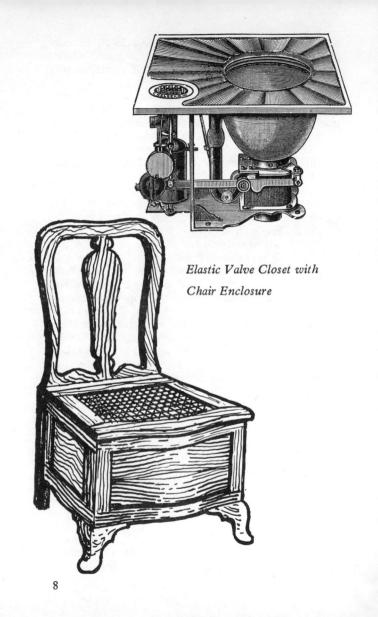

Elastic Valve Closet with
Chair Enclosure

survive. The river was the main attraction. Before the construction of the Embankment in 1871 set up what was in effect a barrier between the buildings and the river, the river front blended in with the houses, the shops and the inns along Cheyne Walk. One could stroll along the bank of the river under stately trees, past rowboats and sailing craft drawn up on the sandy strip at the water's edge and if of ample means have one's own landing stage from which to set off downstream to Westminster on business or up the river for a pleasure trip.

It was no wonder that Chelsea proved such a magnet to artists and writers, such as Turner, Leigh Hunt and Thomas Carlyle. There is no record of Crapper clearing a clogged drain for Swinburne, repairing a kitchen tap for Whistler or installing a bidet for Christina Rossetti. But he might very well have done so, for they all lived a mere stone's throw from him and when they had the need of a plumber there is no reason why a maidservant should not have been dispatched to fetch Crapper.

In 1861, after thirteen years of industrious work as a plumber, Crapper was able to set up in business for himself in nearby Marlborough Road as a Sanitary Engineer. The Marlboro' Works of Thomas Crapper & Co. (Telegrams, 'CRAPPER, CHELSEA') were at 50, 52 and 54 Marlborough Road but if you go there now you will see that the site has been swallowed up in a big block of flats, built in the 1930s. And while they were about it they changed the name of Marlborough Road, incorporating it into an extension of Draycott Avenue. The reason for this was that London had more Marlborough Roads than it knew what to do with. The British had been very proud of the Duke of Marlborough and in that era when the city was really spreading out, local boroughs hastened to have at least one street named after the General who had been victor of Blenheim, Ramillies, Oudenarde and Malplaquet.

By the 1930s all the Marlborough Roads, Avenues and Streets were driving the Post Office crazy, so they set about having the name of some of them changed. Today there are still 33 Marlborough streets in London.

Crapper could not have started up in business for himself at a better time. The year 1861 was to be the start of a boom period for plumbers, occasioned by the fact that London had, belatedly, just got its first two main sewers, which in the next four years were to be extended to a network of 83 miles of large intercepting sewers. Crapper, along with all the other plumbers, was happily inundated with work.

His outside staff were kept continually on the go; work poured in to his Marlboro' Works. It was a two-storey stucco building with an archway leading through to the yard and the brass foundry. Work from the foundry went upstairs to the brass finishing department and also upstairs were the offices and showrooms. The storage rooms were at the back of the ground floor, adjoining the shop, with its counter so high that it was at about eye level since in Victorian times customers could apparently not be trusted to keep their thieving fingers off the merchandise when the assistant was out in the back. And the main part of the ground floor (most important as far as we are concerned at any rate) was occupied by the cistern workshop from which such exciting new things were to emerge.

Chapter 3

'Pull and Let Go' Is Born

Undoubtedly Thomas Crapper's greatest, most lasting contribution was the work he did in developing the modern W.C. cistern. He was prompted to turn his attention to this when new regulations came into force following the Government's Metropolis Water Act of 1872.

This piece of legislation had been essential, to clear up a terrible state of confusion. At that time there was not a single authority dealing with the whole of London's water supply as with the Metropolitan Water Board today. Eight separate water companies had the city divided up into sections and this was a terrible headache for the plumbers, since each local company had its own standards and requirements as regards water installations. Whenever their work took them from one area to another – from Chelsea, say, to Lambeth or Central London – plumbers found that they had to conform to different regulations. The Metropolis Water Act unified the whole thing.

And one of the main regulations was aimed at curbing the shocking wastage of water that was going on in the lavatories of the metropolis.

In the old days the water for a flushing toilet was provided from a cistern in which there was a valve at the outlet to the flush pipe. When you pulled the chain it simply lifted up that valve and released the water. In other words you just pulled the plug out. Some people would tie the chain down

11

Crapper's Valveless Water Waste Preventer.

(Patent No. 4,990.)

One moveable part only.

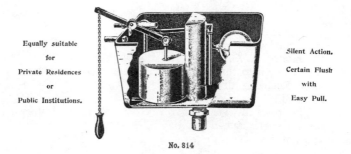

Equally suitable
for
Private Residences
or
Public Institutions.

Silent Action.

Certain Flush
with
Easy Pull.

No. 814

Quick and Powerful Discharge maintained throughout.

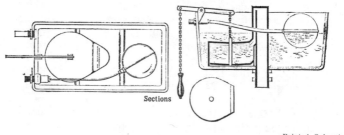

Sections

		Painted	Galvanized
No. 814	2-gallon Crapper's Valveless Water Waste Preventer with Pull...	20/6	26/3
	3-gallon Crapper's Valveless Water	23/6	30/6
	If with Cover ....Extra	1/9	3/-

so that the valve was perpetually open and the water flowing ceaselessly – either because they were too lazy to pull the chain every time or because they were ultra fastidious and wanted to ensure an immaculate flushing of the bowl.

This sort of thing horrified the Board of Trade, which used to be the ministry responsible for our water supply. They envisaged enough people doing it to cause all the reservoirs to dry up, and drought and pestilence could strike the land.

But even worse was the second factor, although it would have seemed to have been of lesser importance. This was the fact that try as they might the makers of the valves could not ensure a snug fit. Each valve would start off watertight but it would not be long in use before it was failing to lodge properly after each flush.

This trickle, multiplied by thousands, was the Board of Trade's big worry. So the call went out for somebody to evolve a 'Water Waste Preventer'.

It would not be absolutely true to say that it was Thomas Crapper who *invented* the Water Waste Preventer, which for many years was to be the technical term for what is now our modern W.C. cistern. It was not as clear-cut as, say, Diesel inventing the engine which bears his name. More than a few plumbers took up the challenge and just as a hundred years previously the work of James Watt and others had culminated in his producing the 'first effective modern steam engine', so it was that there came into being 'Crapper's Valveless Waste Preventer. One Moveable Part only. Certain Flush with Easy Pull. Will Flush When Only Two-third's Full.' In other words he perfected the cistern as toilet users throughout the world know it today.

The trick was to make water flow uphill. If you consult the drawings opposite you will see how this was done. Or if you want to study it in the round, as it were, go to your nearest cistern and lift the cover and you will see precisely the same

13

principle Thomas Crapper and his aides evolved at his Marlboro' Works in Victorian times.

The focal point, where the action is, is the circular chamber to the left of the central pipe leading down eventually to the toilet. At the base of this chamber, which lets water in because it does not come quite to the bottom of the cistern, is a plate attached to a rod which moves it up when the chain is pulled. As the plate moves up it takes with it water which goes up one side of the partition in the central pipe and overflows into the main part of the pipe. The top of the pipe must be above the level of the contents of the cistern, so that there is air there. The water brought up from the chamber displaces this air and this sets up a siphonic action which sucks all the water out of the tank and sends it cascading down the flush pipe with what Crapper described as 'considerable velocity', which is essential for thoroughly cleansing the basin. There can be no leakage because it is impossible for the water to go up and over the partition until somebody pulls it up. Ingenious, isn't it?

After the cistern has emptied, water automatically comes in for the refill. This is done by the float (the big copper ball of Crapper's day, plastic now). As the level lowers during the flush the float comes down and the metal arm to which it is attached opens the inlet from the water supply. The water gushes in and continues to do so until the level has risen to the height at which the float will have brought the metal arm up to the position which turns the inlet off. And the water will sit there, with no wastage, until somebody comes along and pulls the chain again.

It was no use trying to wedge the chain in such a position that you got a perpetual flush. With this new system it just couldn't be done. 'Pull and Let Go' was born.

As with Stephenson and 'The Rocket', say, or the Wright brothers with their aeroplane, Crapper did

14

not bring his Water Waste Preventer to perfection until after many a dummy run.

The testing of toilets was no haphazard business. Crapper had a test panel of five at his Marlboro' Works – a row of five pairs of brackets up on the wall to take the cisterns, a ladder handy for the workmen to fit them on to the flush pipes leading down to the toilets underneath, the whole served by a 200 gallon tank of water up on the roof. This test panel was in constant use, for not only were Crapper and his helpers continually working on new ideas, but all the cisterns they made in their regular range and every one brought in for repair were subjected to stringent tests in this simulation of everyday conditions before he would let them leave the premises.

And it wasn't enough merely to say, 'The water comes down all right, she's fine.' Each toilet would have to be seen to perform an efficient flush.

The obvious difficulty, not to say indelicacy, of making the tests completely true to life was overcome by using a variety of commodities to represent the 'soil' that had to be flushed away. These consisted of such things as apples (surprisingly enough), sponges, cotton waste, grease and 'air vessels'.

The last named, in their simplest form, were crumpled up pieces of paper which, since they encased air and were resistant to water, were difficult to flush. I was shown by a Crapper Old Boy how they evolved a more sophisticated type of air vessel. Taking a piece of paper by one corner he folded it around his fingers to make the sort of cone shaped bag confectioners used to sell loose sweets in for youngsters in the old days before pre-packaging. Also he demonstrated another version like a little Christmas cracker. Such air vessels floating on the water in the lavatory pan were a real challenge to the flushing power of a cistern.

The magazine *Health*, reporting on the toilet testing activities at the Health Exhibition of 1884, told

15

Thomas and his workmen at the test panel
Drawing by Edith Crapper

E.B.CRAPPER

of a super-flush which had completely cleared away:

10 apples averaging 1¾ ins in diameter,

1 flat sponge 4½ ins in diameter,

3 air vessels,

Plumber's 'Smudge' coated over the pan,

4 pieces of paper adhering closely to the soiled surface.

This is a record which has stood till this day and is as unlikely to be eclipsed as Laker's 19 wickets in a Test against Australia.

One can visualise the sense of excitement among these pioneers, having worked and worked to get the bugs out of some revolutionary flushing device, crowding round the toilet as the Old Man stands, chain-pull in hand, ready to give his brain child the big test. And the exuberance of success. There is the incident well known in the trade of one apprentice having his cap snatched off his head and tossed in among the ersatz 'soil' and the triumphant cry of 'It works!' as his cap went down the drain with the greasy apples and sponges.

Edith Crapper tells of visiting the works one day when her father and Crapper were testing a prototype and a little crowd of workmen had gathered at the test panel to see the moment of truth. But on this occasion it wasn't a success. The intake valve was either faulty or had not been fitted properly. At any event when the chain was pulled the valve came partly off its moorings and the full pressure of the 200 gallon tank up on the roof shot out a spray of water which showered over everybody. As an earlier Peter Arno would have said, it was a case of 'Oh well, back to the old drawing board.'

Chapter 4

The Chain That Won't Work

At this point you may feel prompted to say: 'If Thomas Crapper *perfected* the W.C., why is it that so often the chain won't work?'

It would be grossly unfair to impugn the good name of Crapper by saying that the fault lies with him. It is *our* fault, not his.

More than one veteran of the plumbing business has told me that it is an interesting quirk of human nature that the average person will not begrudge money put out on new furniture, kitchen appliances and so forth. But when it comes to the bathroom, they pinch pennies.

How often one has gone into a pub to find it has been given a face-lift, the bar remodelled and refurbished, no expense spared . . . but go out to the Gents and it's a thousand pounds to a gooseberry that it's the same old pull-and-let-go job that was there when grandfather had his first pint as a youth. For some odd reason, many people seem to think that a W.C. cistern was made to last for ever. The point is that it *is* a mechanical appliance. It has working parts. With age, they become inefficient.

Before embarking on the research for this book I thought that The Chain That Won't Work was merely something of that period between the wars, when there were still in use vast numbers of toilets which the Victorians had bequeathed to the world. Older people will remember that families, after a long period of trial and error, evolved complicated rituals for coping with these temperamental loos. Visitors heading for the bathroom were likely to be given complicated instructions on the lines of 'You pull and hold it for a while, then give two quick jerks and it'll flush.'

I thought that the great amount of rebuilding in the post-war years had eliminated much of this problem. But it seems that many people – in mews flats in Kensington, bedsitters in Earl's Court and so on – are still each day confronted with the frustration I remember in my youth – pulling and pulling at a chain and being rewarded only with a dry clanging up above and a mocking trickle down below.

What makes it so annoying is that, in one's ignorance, one assumes that when the chain is pulled it merely pulls out the plug and releases the water. How could anything so simple fail to function?

If you consult the picture on page 12 you will see the key to the problem, the one thing to which all your frustration is traceable. In the drawing of the cistern bottom right you can see clearly the plate at the bottom of the chamber which is attached by a short chain to the metal arm which has at its other end the main chain. Now, we have explained elsewhere how pulling the chain raises this plate and brings enough water up over the partition in the flush pipe to start the siphonic action which empties the cistern and flushes the toilet. This unit is of cast iron and as the years go by it rusts and it also accumulates deposits of lime from the water (like the inside of your kettle). Result of this is that the plate no longer moves freely up and down in the chamber. It jolts up and down and goes askew and only when you're lucky carries up enough water to start the full flow – hence the trickle. But what makes you really infuriated is when the plate gets stuck up at the top and you're pulling away there without any effect on it whatsoever; you're merely taking up the slack of the little bit of chain and then letting it fall loose over and over again.

No. 231

No. 232

Chapter 5
Quiet, Please!

Akin to the problem of the recalcitrant chain is the noise associated with the W.C. A question frequently asked, particularly by women, is: 'Why can't there be a really silent toilet?'

Sensitive females, such as maiden ladies and girls entertaining a new boy friend, know it is no good announcing, 'Excuse me a minute, I'll see if the scones are done', merely to have the sound of the flushing toilet proclaim to everybody in the house, and even out in the garden on a still evening, 'I have been to the loo.'

Crapper had this problem very much in mind and in those pioneer days of the modern toilet he knew that the noise broke down into these four basic components:

(a) The sound of the flush itself, the down-rush of water into the pan;

(b) The gurgle which occurs at the end of the flush;

(c) The hissing sound of the water coming in to refill the cistern;

(d) The noise of water coming under high pressure to the cistern through copper pipes which, to quote an old time plumber, 'ring like a peel of bells'.

When Crapper was confronted with the problem of what to do about it all, this water concerto was really *con brio*. Thanks are due to him for building in silencers which cut down much of the noise and to which modern water closet makers have added little more in the way of noise elemin-ation. Crapper's improvements to what became the 'Marlboro' *Silent* Water Waste Preventer at the turn of the century incorporated two refinements which, if they didn't 'prevent the hissing and the gurgling' as claimed, at least minimised it. Modern inventors have got rid of some of the gushing noise down in the pan, by developing the siphonic trap. But truly Silent still eludes them and it was cause for worry on a Royal level.

At the coronation of our present Queen the organisers of the occasion were concerned, among other things, about the matter of the special needs of the great number of Peers who would be assembled at Westminster Abbey, many of whom were well on in years and not able, with the best will in the world, to stay settled in one place for long periods, as the lengthy ceremony demanded. It was known that the very aptly named 'peer's bladder' would come to their aid, but there was more to it than that. So an additional 'range' of toilets, as they call it in the trade, had to be installed to cope with the expected increased demand.

Then somebody got the frightening thought that at the vital moment of the ceremony, when the whole Abbey would be in hushed silence as the crown was being placed on the Queen's head, there might be one of those terrible coincidences whereby all the toilets happened to be occupied and all their occupants pulled the chain at the same time. Would the strains of that symphonic flush penetrate into the body of the Abbey and create one of the major embarrassments in the history of British Royalty? There was nothing for it but to have a test.

A detachment of Guards from nearby Wellington Barracks was pressed into service, and as technicians borrowed from the BBC, each armed with a deci-bel meter, were stationed at various key points in the main part of the Abbey, the troops were deployed along the long line of toilets. It is not

22

for miners in the pits it was something which a plumber merely encountered from time to time. Or perhaps they just thought they would look silly going down to the basement with a safety-lamp on.

H. A. J. Lamb, writing in *The Architects' Journal*, quotes an item from a contemporary piece on the hazards of the plumber early last century: 'Two workmen had a narrow escape for their lives, for, upon opening the flags, one of them, bowing down to examine the shaft, was suddenly surrounded by flame from a lighted candle in his hand. There was an explosion which split a water bucket, stunned the men, and shut the door, which was half open, with a great noise.'

An incident in modern times and on a much larger scale forcibly illustrates the perils the old time plumbers faced in this regard. It was what was to become known as the Palestinian Explosion of 1947. I was told about it by an acquaintance of mine who had been with the Medical Corps attached to the First Infantry Division during the Palestine troubles just after World War II. In his camp there was a permanent latrine which might well have been termed historic, since it had obviously done service for several wars. It was a huge concrete-encased, circular pit, the concrete having been continued up above ground to form a parapet across which the seat-boards were laid. The occupants, divided by wooden partitions, sat outward looking in the style of the garderobes still to be seen in old English castles. The edifice was topped with a corrugated iron roof.

At that time DDT had just come into being and the Disinfestation Officer decided to put some of this to work on the foul smelling accumulation down below. The DDT was in solution with a paraffin base and he poured in a couple of gallons of this. There had been complaints from late night users of the latrine about the disconcerting sound of rats down there, so he thought it would be a good idea

known what form of drill was evolved for this unique exercise. It was probably something like: 'At the command, "Chains – PULL" . . . wait for it!' Anyway, as the order rang out all the toilets were flushed simultaneously and the good news from inside the Abbey was that nothing could be heard of the noises off. The heading *Time* magazine put on their story about it was 'Royal Flush'.

No. 792

No. 792.   Seat Pedestals   ...   per pair   6/9

Galvanized do.   ...   „   11/9

No. 794

Chapter 6

Hazards of the Victorian Plumber

Elizabeth Longford in her admirable book *Victoria R.I.* wrote: 'In 1858, when the Queen and Prince Albert had attempted a short pleasure cruise on the Thames its malodorous waters drove them back to land within a few minutes. That summer a prolonged wave of heat and drought exposed its banks, rotten with the sewage of an overgrown, undrained city. Because of the stench, Parliament had to rise early.'

Not only were the politicians of Queen Victoria's day denied the pleasure of a stroll and a chat on the terrace of the Houses of Parliament, as so much enjoyed by MPs and Peers today, but so penetrating was the effluvium from the river that there was debate as to whether the building itself was tenable as a seat of government. So frequently did Parliament have to rise early when there was an off-river breeze that it was a question whether they could get through their business by the end of the curtailed session and it was felt that it would become necessary to move to another site.

The River Cam, like the Thames, was for many years not the best of places for a pleasure outing. Gwen Raverat, in her book about her Cambridge childhood, *Period Piece*, wrote: 'I can remember the smell very well, for all the sewage went into the river, till the town was at last properly drained, when I was about ten years old. There is a tale of Queen Victoria being shown over Trinity by the

Master, Dr Whewell, and saying, as she look[ed] over the bridge: "What are all those pieces [of paper] floating down the river?" To which, wi[th] presence of mind, he replied: "Those, ma['am, are] notices that bathing is forbidden." '

In matters of sanitation the early Victor[ians] progressed little from the darkest days w[hen] filth of the home was emptied into the str[eet;] river, stream or ditch was regarded as a goo[d] disposal area and in the early years of [her] reign the Fleet River (now Farringdon Stre[et]) was still an open sewer, not to be covered [until] 1844. When no stream was handy, as in th[e case] of Sherbourne Lane in the City which skirt[ed the] bourne of sweet water' and soon came to [be known] as Shiteburn Lane, the cesspool was still th[e only] way to cope with the sanitation problem. I[n 1844] fewer than fifty-three overflowing cess[pools were] found under Windsor Castle. There was t[he tale of] one titled host, at the front door of hi[s stately] mansion to greet the arrival of a coach ful[l of week-] end guests, being called upon to watch [in horror] as the driveway subsided and they were e[ngulfed in] an overflowing cesspool. And, sad to rela[te,] not without loss of life.

Such was the atmosphere in which [Thomas] Crapper had started up in work, wher[e to be a] plumber was, to say the least, a hazardous [calling.] Delving around among grossly inefficient [sewers and] drains that at best were poorly ventilat[ed, and at] worst not ventilated at all, you laid you[rself open] constantly to being laid low by one of the [two main] dangers – explosion or dread disease.

The fumes that wafted up from faulty [drains and] old cesspools were not only foul and de[bilitating to] those who lived over them. They were als[o lethal.]

Sir Humphry Davy invented the safe[ty lamp in] 1815 but although miners were quick to [use it,] plumbers did not follow suit. Perhaps th[is was] whereas explosive gases were an ever pre[sent]

to complete the operation by burning them out. He threw in a lighted match and was not prepared for what happened. When the paraffin caught alight it ignited the mephitic fumes and the force of the explosion was such that the concrete parapet split open like the petals of a flower and the tin roof was sent a considerable distance into the air. And the aftermath of this atomic-like detonation was that fall-out, in the form of toilet paper, was still drifting down out of the sky four hours later.

When I asked what had happened to the sanitary officer, my ex-Medical Corps friend said: 'He lost his eyebrows and was in for shock.'

Undoubtedly he had escaped with his life merely because it was out in the open and not in a confined space. I mention elsewhere how at least one of Crapper's associates was to wind up in hospital as a result of similar foolhardiness with a naked flame but Crapper himself appears to have escaped any such misadventure. But what did lay him low was that other major hazard – disease.

Two highly infectious diseases which were rampant at the time and which were almost entirely traceable to bad sanitation were typhoid and smallpox. Queen Victoria's beloved Albert died of typhoid in 1861 and son Edward, Prince of Wales was lucky to survive an attack ten years later. In 1870 one in every 3,000 of the British population died of typhoid; by 1900 the death rate was still as high as one in 5,000. (For comparison: Britain's death toll on the roads today is one in 7,000.)

Crapper came down with smallpox in 1887 while waiting for a train at Streatham Hill station. It was evening and as he lay on a bench in the waiting room shivering and aching all over with a violent temperature he was too ill to be embarrassed by the fact that the few people who were about mistook him for a drunk, of which in those days of the 'gin-palaces' there were far more to be seen than today. It was not for some time that his true state was realised and he was rushed home, where his good wife Maria nursed him back to health, rarely leaving his bedside for three weeks.

An interesting thing to note is that Crapper and other plumbers of his day laid themselves open to infection not only by direct contact during the course of their work. Crapper, perforce, spent a great deal of time flushing toilets, and one learns from a recent issue of *Lancet* that the water closet is a fine example of an aerosol.

Not visible to the naked eye, the spray sent out by a toilet has given hospitals cause for concern and in 1966 *The Lancet* carried a report on tests made in connection with the fact that 'flushing a wash-down water closet produces a bacterial aerosol'.

Toilets of various designs were set up 'in exactly the same position in the same room' and the water contained in each was contaminated with 'an overnight broth culture of *Esch. Coli*, carefully poured from a height of 15 cm. to avoid splashing.' The toilets were flushed and droplets in the spray that shot out from them were gathered on 'slit-samplers'. Appropriately enough, these tests were in the capable hands of the director of the Middlesbrough Public Health Laboratory, Dr Blowers.

After the slit-samplers were incubated for 24 hours, not only was the disturbing fact revealed that the average wash-down closet had sent out 37.5 colonies of bacterial contamination per 100 cubic feet. Tests made with the seat cover down resulted in 'a rather surprising finding'. Far from the seat cover blocking the spray, the number of bacterial colonies jumped from 37.5 to 46.9. Apparently the spray was forced out through the gap the rubber buffers create between the seat and the rim of the pan and this made the toilet into a much more efficient aerosol, like turning the nozzle of a hose to 'jet stream'.

All of which makes a nice new thing for anyone with a contamination complex to worry about.

Chapter 7

'By Royal Appointment...'

Being asked to do the drains and bathroom installations at Sandringham was the high-water mark of Crapper's life. It was the start of what was to culminate in the firm being granted four Royal Warrants over a period of half a century.

If anyone had any doubts about Thos. Crapper being the Royal Plumber they were at once dispelled when his Marlboro' Works came into view. The dominant, eye-catching feature of the two-storey building at 52 Marlborough Road, Chelsea, was the 6 x 4 ft. Royal Crest which crowned the stucco façade. Under the glistening blue, red and gilt coat-of-arms, in lettering a foot high which ran the full length of the rooftop was the legend: By Appointment MANUFACTURING SANITARY ENGINEERS.

In this book there is an old photograph of Thos. standing in front of his works in billycock hat and frock coat with his office staff and workmen in long white aprons, bowler-hatted except for the apprentices in cloth caps. The frontage of three display windows and an archway leading to the yard at the back is identified as the premises of 'T. CRAPPER, Lead Merchant, Brass-Founder and Manufacturer of Patent Regulator Water Closets.' Although the merchandise in the windows and arrayed along the pavement included a bath tub, two sinks and various lengths of pipe, there are 27 lavatory pans on display. There is no question that

Crapper's heart was in the toilet.

Inside the building there was no likelihood of the visitor missing the Royal association, if by odd chance he had not noticed the regal display on the rooftop. On the stairs leading up to the showroom were the four framed Royal Warrants, granted by Edward VII when Prince of Wales and when he ascended the throne and George V as Prince of Wales and as King. The frames were elegant affairs made by a firm which specialised in that type of thing. They were of gilt, topped, in the case of the King, with a crown on a cushion built into the frame itself, and for each Prince of Wales the *fleur de lis*. Unfortunately they don't exist any more, since it is a stipulation from the Palace that holders of Royal Warrants must destroy them once the warrant has expired either through the death of the Royal personage or of the warrant holder. It is not generally known that Royal Warrants are in fact granted to persons, not to firms, and any trade or business company cannot keep the 'By Appointment' regalia on display once the warrant holder is dead, although the Palace does allow six months grace to get everything off show. So strict is this rule to avoid abuses of Royal Warrants that firms cannot even keep them as souvenirs, to display in a private museum. The wording, printed in blue script type, was of delightful Englishness:

These are to certify that I have appointed
Mr Thomas Crapper, trading under the
style of T. Crapper and Company, Chelsea
into the Place and Quality of
Sanitary Engineers to His Majesty.

To have, hold, exercise and enjoy the said Place together with all Rights, Profits, Privileges and Advantages thereunto belonging.

This appointment is personal and does not

> *extend to any further member of the firm.*
> *And for so doing, this shall be your Warrant.*
> > *Given under my hand,*
> > *(S.) Lathom*
> > *Lord Chamberlain*

The giant coat of arms would have been a prize indeed for any Crapper souvenir hunter, although its very size would have made it dominate any but the most spacious of living-rooms. However, the workmen called in to remove it let it drop when lowering it down and being of plaster it shattered on the pavement.

A sad and unseemly end to a proud association with our monarchs, which thankfully old Thomas was not to witness.

Not all the orders Crapper got from Edward VII were for the Palace. There was an item which consisted of a W.C. enclosure in the form of an attractive armchair upholstered in velvet which was delivered to and installed at an address in West Hampstead.

Lily Langtry was born in 1853, daughter of the Dean of Jersey, the Rev. W. C. le Breton. Such were her good looks that when she came to London to pursue a career on the stage she became known as 'The Jersey Lily'. But although she was capable of a fairly good Rosalind her main success came not in the theatre but in London society. Her entrée was through her marriage to a wealthy Irishman named Edward Langtry, who had a racing stable and all the other trappings of a leading light in the social set. It was not long before she had caught the eye of Edward, Prince of Wales, and they became friends.

When Edward Langtry passed out of the picture the Jersey Lily was installed in a house on the fringe of St John's Wood, that London area which had come into being as the world's first planned 'garden suburb' after the opening of nearby Regent's Park in 1830 and which had become a fashionable place for the secluded residence of the mistresses of eminent gentlemen, Royal and otherwise.

A clue as to why Miss Langtry's new home was not actually in St John's Wood but adjoining it lies in the name of the street. For her to move into Alexandra Road was doubtless regarded as too good an 'in' joke to pass up. There were in fact allusions in the public prints of the day to such a choice being lacking in good taste, especially since the Prince at that time had been receiving letters from Queen Victoria to the effect that he should 'mend his ways'. Indeed, as late as 1968, when Lily Langtry's home was about to be pulled down to make way for re-development, this touchy point was still very much alive.

Leighton House, where 'the Prince of Wales was a frequent visitor to tea', had floors and fireplaces of marble, stained glass windows and drapes of yellow damask. The 'removable armchair' which fitted over the toilet in the bathroom next to the master bedroom was made by Crapper's joiner, Charlie Vivian. It was a more elaborate, custom-built version of Crapper's standard 'Closet Chair' (No. 1658) and was part of the service he provided – 'Any Design of Closet Chair Made to Order'. It cost £9 12s. 6d. Today a specially built upholstered armchair of that type (with a permanent seat, of course, not the tip-up variety) would cost something in the region of £70. A friend who had seen the Langtry armchair was asked what colour the velvet was and replied: 'Royal blue, of course.'

Chapter 8
Sandringham Days

Sandringham House was where it all began as far as Crapper the Royal Plumber was concerned and visitors today can still observe that he had indeed been at work there. On the pathway leading up from the rockery to the house there is not merely one but two Crapper manhole covers.

It is clear that when he had been summoned to Sandringham it had been borne very much in mind that there would have been nothing worse than a member of the Royal Family escorting guests on an evening stroll down to the rockery to take in the fragrance of the freesias and the night scented stock and being greeted by whiffs of sewer gas. The Crapper manhole covers would ensure that it would be impossible for that to happen. When put in place it was no happy-go-lucky fit. Although they were not actually gauged to a thou, their manufacture was a precision operation and even the most persevering sewer fume had no chance of getting up through the snug lodgement.

Sandringham is open to the public and in the summer thousands of visitors take the opportunity to stroll along the gravel paths that skirt the beautifully manicured lawns and see the gardens and rockeries in among the clusters of trees on the huge estate. But the house itself, the big mansion on high ground dominating the whole scene, they may view only from the outside.

The Prince Consort had been in the process of

buying Sandringham in 1861 as a twenty-first birthday present for Edward, Prince of Wales, when he died before the final papers could be put through. Despite the raised eyebrows about the price – a staggering £220,000 – Queen Victoria insisted that since it was her husband's wish, the deal should go through. After all, the money was not coming from the taxpayer but from accumulated funds of the Duchy of Cornwall, the hereditary perquisite of the heir-apparent.

In 1862 Edward finalised the purchase of what was then called Sandringham Hall. The enlarging of it took several years. Eventually it was almost completely demolished and by 1870 the new Sandringham House had taken its place. With various additions over the years it now stands a many-gabled three-storey red brick and stone country home with a garden frontage some two football fields wide. Queen Victoria had described it as 'a handsome newly built Elizabethan building'. More recently it has been alluded to as 'looking like an Edwardian hotel'. It is now generally agreed that it was designed during 'an unfortunate period of English architecture'.

By the middle 1880s the drains and general sanitary arrangements at Sandringham had not merely deteriorated with the years. They had been rendered absolutely antiquated by the great advances made by Crapper and his contemporaries in the field of sanitation. When it was decided to fix things up it was not just a matter of calling in the local plumber. The house itself, neighbouring York Cottage and all the other outbuildings on the estate formed such a vast complex that it was necessary to call in a firm of civil engineers from London to do a survey. Their 'Report on the Drainage of Sandringham House' of February 18, 1886, occupied 42 pages of meticulously executed handwriting.

Work had apparently been put in hand at once, for later in the same year there was a beautifully bound

34

volume in red leather with 'Drainage of Sandringham House, Book of Reference' in gilt lettering on the cover. Still in existence are several of these books of reference, going through to as late as 1909. Every bit of work was itemised and numbered and the name Crapper catches one's eye on page after page of details of flushing tanks, disconnecting traps, inspection chambers and bathroom fittings. Obviously Mr Roger Field, head of the civil engineering firm in Victoria Street, S.W., had not needed to look further than Thomas Crapper in nearby Chelsea for the key man to put his recommendations into action.

There are more than thirty toilets in Sandringham but recently there was a massive overhaul of the bathroom arrangements and, sadly for anyone interested in the work of Thos. Crapper, many of his original installations were deemed to have outlived their usefulness after some seventy years of faithful service to our own and visiting Royalty.

The Queen's amenities had been completely modernised, which is as should be. The bathroom of her Lady-in-Waiting, however, is a gem. Set in a huge slab of white marble are three washbasins in descending order of size. The largest is labelled in leaded lettering HEAD & FACE ONLY, the next one HANDS and the smallest TEETH. Why the Victorians felt this division of labour was necessary it is hard to say. Each basin has its own two taps and there is no danger of losing the plugs, simply because there aren't any. The basins are of the tip-up variety, each on a central pivot so that you merely lift it up by the little lip in the front and tip out the contents after use. There were products of Jennings, a contemporary of Crapper who made some notable contributions in the field of lavatory pottery. But the adjoining bath *was* a Crapper product. Roomy enough to invite a friend, it stands on 'ball-and-claw' feet, and it was a Crapper 'Kenwhar', a composite name made up from *Ken*sington, adjoining Crapper's works, and *Whar*am, his then managing director.

The toilet likewise is a Crapper and it confirms what had been said by Robson Barrett, who had been the last managing director at Crapper's. 'At Sandringham all the original Crapper toilets were installed with cedar seats and enclosures. Cedar in those days was even more expensive than mahogany or walnut. It had the advantage of being warm. It required no varnishing. And I remember reading somewhere that it was "subtlely aromatic", which was a great thing in its favour for that sort of use.'

The one jarring note in this Lady-in-Waiting's bathroom which is a varitable mini-museum of Victorian plumbing is the cistern above the cedar-encased toilet. It is a new black plastic affair. Clearly it has replaced the 'T. Crapper Chelsea' cistern which would have completed the picture but which obviously had been one of those which had been carted away when it had covered its span of three score years and ten.

However, various guest rooms still have fine examples of water closet enclosures of mahogany with cane back and seat cover (Crapper's No. 1658) and the servants' quarters, in the older part of the great rambling house, is a happy hunting ground for anyone who would wish to see still in use the old original Water Waste Preventers.

In one of the servants' work rooms is a Crapper 'Cecil' Slop Sink with Flushing Rim. The cistern was the same as that used by Crapper for toilets, the chain pull showing that nice touch of class distinction by being a plain oval ring (price in the 1880s, including the chain: 4d.), rather than a Crown Derby or Cream and Gold Fluted China handle as supplied for their superiors.

On the wall by the slop sink is a charming piece of Victoriana which reminds one of the notices that seaside landladies put in bathrooms telling guests what they must remember to do and refrain from doing. This one read: 'NOTICE. This sink must be thoroughly flushed after each use.' However, it was not so much what it says but how it is said. Beautifully lettered with a quill pen and enclosed behind glass in an attractive little frame, it had undoubtedly hung there from the day the new contrivance had been installed in the 1880s.

Nearby are two toilets which escaped the discarding of many of Sandringham's older bathroom installations. They are genuine originals of the *Marlboro*, each with 'T. Crapper & Co, Chelsea, London' up there on the cistern – Old Eight-One-Five as Crapper's employees affectionately termed it. They are genuine antiques, rightfully preserved, if only by chance.

In the main body of the house no male visitor can fail to be intrigued by the urinals at Sandringham. It is not so much that there should be urinals in what, after all, is a family country home. The two in the washroom adjoining the billards room are unusual. At the base of each is a footplate and when you stand on it there is an automatic flush. Today this is a novelty; most certainly guests at Sandringham in Victorian times must have been greatly impressed. The urinals themselves are undoubtedly Crapper installations but anyone interested in finding out who evolved the fascinating footplate flushers will look in vain for a credit line on them. However, one would like to think that they were a further example of the inventiveness of our man.

Chapter 9

A Twyford Cameo

Thomas Crapper worked very closely with another Thomas – Twyford of Stoke-on-Trent. The Twyford firm, it is almost unnecessary to point out, was, and still is, at the forefront of the lavatory pan world. The potters of Stoke-on-Trent needed Crapper, to incorporate their product in what Victorian plumbers called a Combination, in other words the whole toilet unit installed in a house or other premises. And Twyfords showed their gratitude to Crapper each Christmas, when Thomas, his nephew George and managing director Wharam would each receive the gift of a 60 lb. chest of tea. They did things in style in Victorian times.

I went up to Stoke, or to be more accurate neighbouring Hanley, to visit the Twyford potteries and when I was being shown over the works there was one section where I was to be transported back to the old days of Crapper and his test panel of toilets (see Chapter 3). Twyford's, too, have a test panel of four. By no means is toilet testing a thing of the past. H. F. H. Barclay, their managing director, and E. S. Ellis, the designs manager, explained to me the workings of their pride and joy – the double-trap siphonic lavatory pan. It differs from the usual water closet procedure in that the pan is flushed clean not so much from the force of the water coming down from the cistern but by the spectacular gush of suction when the siphonic action is set in motion in the trap underneath as you start the flush.

While we were talking Mr Barclay went to the roll of paper by this one of the four toilets in a row and began drawing out a length of it.

'I think we'll do a seven,' he said, and I didn't understand what he meant until I realised that he was counting off the number of sheets on his strip of toilet paper.

He neatly snipped off the required length and put one end of it in the water in the pan. Then, standing back with it stretched between him and the toilet to give the full dramatic effect, he got Mr Ellis to switch the lever. The toilet paper streamer shot out of his hands and snaked down into the trap to disappear for evermore. I must say that it was impressive. He said he could do a nine or even a ten, which must be even more so.

By way of an encore, Mr E. S. Ellis produced three ping-pong balls and tossed them into the adjoining toilet.

'Most difficult thing in the world to flush,' he said, as I was reminded of Crapper and his air vessels. 'Try to flush them in your ordinary washdown closet, try again and again, and they'll come up smiling every time. But just watch.'

He pressed the lever and whoosh! the three ping-pong balls disappeared.

'That's your siphonic action,' he said.

Of course, the necessity to flush ping-pong balls down the toilet rarely crops up in everyday life but the owner of a double-trap 'siphonic' can take comfort from the fact that he has the toilet for the job when it arises.

I am sure that they did not feel that they looked incongruous, in their blue executive's suits on their way to sherry in the boardroom prior to luncheon, flushing lengths of TP down the loo. After all, it is their work. It is what they are interested in.

Twyford's had been pioneers in bringing the toilet out into the open.

It had always been boxed in with a wooden enclosure, right from the time of Elizabeth I's godson Sir John Harington, poet and satirist, who evolved a fanciful flushing W.C. as long ago as 1596. This habit which persisted of hiding the 'unmentionable' beneath a wooden surround naturally satisfied the Victorians' sense of delicacy. But it had odious drawbacks.

In 1885 Twyford's launched their evolutionary *Unitas*, which they claim to be the first example of what came to be known as a 'pedestal closet', standing alone in full view. They advertised it as 'The Perfection of Cleanliness', since 'no Wood Fittings are required except a hinged seat, which, being raised, free access can be had to all parts of the Basin and Trap, so that everything about the Closet can be easily kept clean.'

This coincided with Crapper's pioneer work with the cistern, for now all the mechanical bits and pieces that had been linked with the pan for flushing purposes and for collecting filth were now moved up to the tank above, well out of the way. Instead of the toilet being a smell exuding single unit, it became two well separated entities – the pan standing by itself and the 'works' high up on the wall with the water supply. Twyford the potter devoted his attention to the former; Crapper the engineer looked after the latter. Which was the start of a happy and prosperous association between the two.

The Angel Hotel in Doncaster were among the first proud possessors of a *Unitas* and the management were even more proud when Queen Victoria on a visit to the town made use of it. Wishing to capitalise on this, they were placed in rather a quandary. Numerous inns around the country displayed notices to the effect that 'Queen Elizabeth Slept Here' but they could not very well make capital out of Queen Victoria's brief visit by erecting a sign on something the same lines. However, it did not become necessary. Word of mouth, as ever,

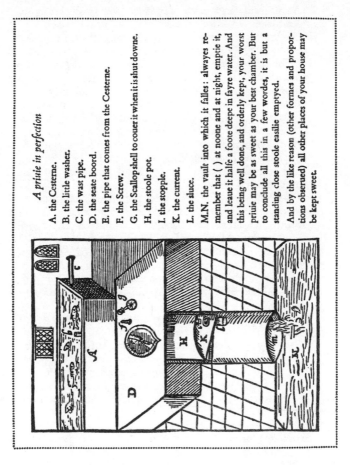

A priuie in perfection

A. the Cesterne.

B. the litle washer.

C. the wast pipe.

D. the seate boord.

E. the pipe that comes from the Cesterne.

F. the Screw.

G. the Scallopshell to couer it when it is shut downe.

H. the stoole pot.

I. the stopple.

K. the current.

L. the sluce.

M.N. the vault into which it falles: alwayes remember that ( ) at noone and at night, emptie it, and leaue it halfe a foote deepe in fayre water. And this being well done, and orderly kept, your worst priuie may be as sweet as your best chamber. But to conclude all this in a few wordes, it is but a standing close stoole easilie emptyed.

And by the like reason (other formes and proportions obserued) all other places of your house may be kept sweet.

Sir John Harington's W.C. of 1596

41

proving the best form of advertising, the news soon got around and each day saw the good ladies of Doncaster making their way to the Angel and queuing up to use the same toilet that had been graced by their beloved Queen.

Having had the courage to bring the toilet out into the open, the Victorians saw to it that it was a thing of beauty. Or perhaps it was that they were rather self-conscious about it and felt that it should be camouflaged. At any rate all sorts of designs were glazed into the pottery. Some had 'Raised Ornamentation'. Others went as far as to incorporate statuary, as with the Pedestal Lion Closet and The Dolphin, with animals bearing the weight of the bowl on their shoulders.

The Willow Pattern was popular. A big seller, especially in the Windsor area, where examples are still to be seen, had Windsor Castle in the bowl, which today might seem to be an odd sort of tribute to the Sovereign. Motifs in the manner of a Grecian frieze were often used.

But by far the most popular were floral designs, such as the Blue Magnolia, the Acanthus, Mulberry Peach and the Bouquet.

However, at the turn of the century with the death of Queen Victoria the ornamental loo went the way of other of the more attractive aspects of the era that bears her name. The toilet became a functional white and remained so until manufacturers in the post-war years varied it by turning to pastel shades.

Twyfords now carry five colours, ranging from pink to turquoise.

'Pink is by far the most popular colour,' the managing director told me.

'Why is that?' I asked.

'It's a warm colour. Second is sky blue, followed by primrose.' Then he added with a smile. 'But our new colour Pampas – the colour of pampas grass – is coming up fast on the charts.'

The other interesting thing that emerged from

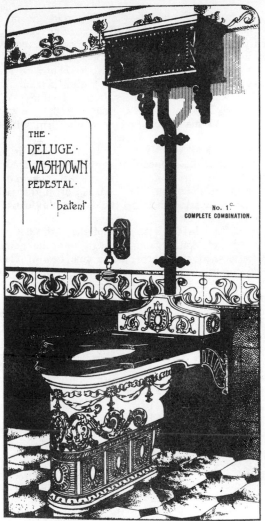

THE ·
DELUGE ·
WASHDOWN
PEDESTAL ·

· Patent ·

No. 1ᶜ
COMPLETE COMBINATION.

*An early Twyford W.C.
with siphonic cistern*

my visit to Twyfords was that they told me of a book which had been published some time ago and which dealt with their early history. I was extremely lucky in getting from Edith Crapper, old Thomas's grand-niece, a copy of *Twyfords: A Chapter in the History of Pottery* inscribed with a quill pen on the flylead, 'With Compliments from Thomas W. Twyford, to Mr T. Crapper, May 31, 1898,' a gift from the then managing director of the firm which we have seen worked so closely with old Thos. in the pioneer days of the toilet. In the book it is admitted quite unashamedly that the establishing of Twyfords was the outcome of one of the most fascinating exploits in the annals of industrial espionage centuries before the term 'industrial espionage' was coined.

It seems that in the latter part of the 17th century two potters of Amsterdam, brothers with the name of Elers, leaned that there was a fine bed of clay suitable for pottery in the vicinity of Burslem in Staffordshire. They emigrated and set up a pottery at Bradwell in 1690. Being of aristocratic stock they felt superior to the uncultured locals and kept themselves very much to themselves.

In those days before the taking out of patents became a reliable way to safeguard your interests it was necessary to be very security conscious. The security precautions of the Elers brothers took a novel form, to say the least. They recruited their servants and workmen from the dullest and most stupid people they could find. They built up what was to be the finest collections of morons, imbeciles and half-wits to be found outside a mental institution. This lowered the general efficiency but it was more than compensated by the fact that since most of their employees didn't even know what they were doing, let alone memorising secret formulae and processes, there was no fear of any of them selling information to rival concerns.

Two local potters, Josiah Twyford and John

Pedestal Lion Closet,
1896

The Sultan with
decoration, 1896

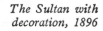

Improved Aeneas with
raised ornamentation

Dolphin, 1880

Astbury, regarded this ingenious form of security as a real challenge. Twyford was the first to get employment at the Elers works. 'He relied upon a carelessness and indifference of manner that went without challenge. It was not a question of assuming a virtue thought he might have it not, but of sustaining the disguise of shrewd intelligence under the aspect of doltish stupidity.'

But Astbury's incursion into securely guarded Elers territory was more spectacular. 'No trained actor could have more thoroughly realised an assumption of character than Astbury. The man employed to turn the thrower's wheel in the pottery was an idiot. Taking his cue from him, Astbury assumed both the garb and antics of an imbecile and is mentioned by the historians as more of a hero than Twyford, who went no further than the assumption of indifference to the operations going on about him, while Astbury suffered bodily in the cause. He submitted to the kicks and cuffs of masters and men with an uncomplaining meekness. He ate his frugal dinner and submitted to the drudgery of his work with a never varying aspect of imbecility. But in his other character of investigator he was wide awake.'

He was fortunate in being moved about to various departments and in the guise of idiot he had to have any function he undertook explained to him over and over again before, supposedly, he could master it properly so that in effect his employers were generously co-operating with him in getting their secrets firmly imprinted on his mind.

Astbury's portrayal was not a single *tour de force,* of the Oscar-winning type. It was more difficult than that. It was a sustained performance over a period of two years.

'By then he came to the conclusion that he had no more to learn in the Elers manufactory, and he availed himself of a real or feigned sickness to remain at home. To prevent any person from visiting

46

him, his sickness was reported to be malignant; and he was thus enabled to perfect and complete the memoranda of his experience and observations and formulate his plans for the future.'

On his return to the pottery Messrs Elers found that not only had his stay at home effected a complete cure but in some odd way his illness had restored him to sanity. They immediately fired him.

'Without suspicion that he possessed a knowledge of all their operations, they soon, however, had the mortification to discover that they were no longer the only persons who could make the pottery they had introduced into the district. Disturbed, if not disgusted, with the inquisitiveness of Burslem potters they packed up shop and departed.'

Both Astbury and Twyford began business on their own account and when eventually the firm Josiah Twyford founded switched from general pottery to specialising in sanitary ware they did so to such good purpose that today theirs is a household name in the lavatory world.

Chapter 10

Inventor at Work

Crapper's development of the Valveless Water Waste Preventer was no mere flush in the pan. He went on to other things. His inventiveness knew no bounds.

His grave concern over the state of the house drains in the London of his era caused him to turn his mind to Ventilation.

As we have seen, the Dark Ages stretched well into the last century as far as the toilet was concerned and one of the worst aspects of that murky period was the smell. Nobody was yet very persistent about taking a bath, so one can imagine the body odour situation. But it did not really matter, for it was as nothing compared to the stench from the spectacularly inefficient drains. It was necessary to evolve smell-thwarters and a typical one was half a pomegranate stuffed with cloves, the forerunner of airwick. But it was not merely for one room in the house, for the smell was all-pervading. And it was potent. People chatting in the parlour would keel over in mid-sentence. The servants got the worst of it, since they had it at first hand, living and working as they did down among the basement drains. It was the main single factor causing them to indulge in gin to the extent they did in Hogarth's day and one can hardly blame them.

When the smell became just *too* much even for the family upstairs in the relatively rarified atmosphere, the plumbers were brought in and they had an ingenious method of locating the leaks in the

Sanitation in the Middle Ages,
from an old woodcut

household drainage. They would pump smoke down the toilet and then knew that the place to go to work would be where it came out through cracks in the kitchen flagstones and elsewhere. Various firms had what were called Smoke Drain Testers on the market and Crapper's early catalogues carried Air Pump and Smoke Generating Machines for this purpose, 'With Specially prepared Smoke Material (oiled Cotton Waste) to smoulder in machine, per cwt, 17s.' And there were smaller, single-action testers such as the 'Wilkinson Drain Grenade' and 'Kemp's Drain Rocket (8s. 6d. per dozen)'. But one drawback to these was that if the drains were in really bad shape there were so many unofficial outlets that a house would become completely full of smoke and would have to be evacuated until it dispersed. Someone then got the bright idea of using oil of peppermint instead, and the Peppermint Test became the vogue. This being a nasal rather than visual form of detection it was not as quickly efficient as smoke. It required an expert to sort out the smell of peppermint from all the other effluvia, whereas any fool can see a puff of smoke. But the plumbers and sanitary inspectors of those days were perforce experts in this regard. They could nose out any specific odour – ammoniacal, excremental, putrefactive or what have you – with the same practiced skill of a symphony conductor catching an E-flat instead of an F-sharp in a full-blooded crescendo.

But Crapper felt that they shouldn't have to be skilled in this field, that people shouldn't constantly be exposed to foul odours coming up from below. So, through the 1880s we see a batch of patents registered under the title 'T. Crapper, Ventilation of House Drains', and these culminated in what is regarded in the sanitary world as vying in importance with his development of the W.C. cistern – his patent No. 10, 332, July 17, 1888 – 'Disconnecting Traps for Safety Purposes'. With this he went right to the source of the whole trouble, for as

Patent Air Pump and Smoke Generating Machine,

For Testing Drains.

No. 1000.

Kemp's Patent Drain Testers.

Per Dozen, 10,6.

Smoke Rockets.

8,6 per dozen.

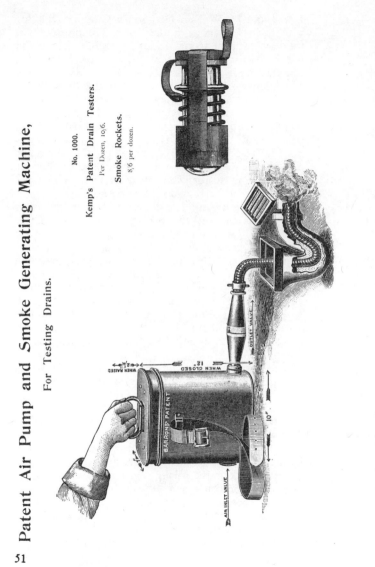

BARRONS' PATENT

AIR INLET VALVE

OUTLET VALVE

WHEN CLOSED

WHEN RAISED

I was told by William Gregory, a Crapper Old Boy who had 44 years with the firm: 'Disconnecting is the key to sanitation.' In other words, if we did not have disconnection now we would be dropping like flies from sewer fumes and be knee deep in rats.

You may not be aware of it but the drains from your house, like any other place where people live or work, do not go straight out to the mains. They go first through your own mini sewage disposal system. All the drains rendezvous at your disconnecting trap, which is a piece of the drainpipe manufacturer's art which I defy anyone to describe adequately in words. One of these from a Crapper catalogue is to be seen on the opposite page.

The waste from your house enters at the open lip of the trap at the left, goes down under the bend and on through at the right to the main drains. There is always water lodged across the bend of the pipe and this forms a perfect seal, preventing the sewer gases from coming up. It is the same sort of water seal to be seen at the bottom of a toilet (the bend the Harpic people have devoted so much thought to) and the kink in the waste pipe under a wash basin and the kitchen sink – all of them added safeguards against fumes coming up into the house.

When I was investigating yet another field to which Crapper directed his inventiveness, I was reminded of something that happened to Eric Coates, the composer. He used to tell of being in a public lavatory and the man next to him was whistling his *Knightsbridge March*. But he kept sounding a wrong note and this so jarred on Coates that he could not help but turn to him to put him on the right lines. But the man took it the wrong way and thought that he was being accosted and it was only after an unfortunate scene and acute embarrassment to Coates that he was able to convince everybody that his motives were perfectly innocent.

I felt I was running the same sort of danger

Patent Disconnecting Trap.

(No. 10,332.)

Registered Design No. 105,149.

Registered Trade Mark No. 81,187, "The Improved Kenon, Thomas Crapper & Co."

Advantages:—Provision at upper part of Trap for discharging into sewer any accumulation caused by accidental stoppage.

Easy access for sweeping purposes, by means of a suitable brass cap with screw.

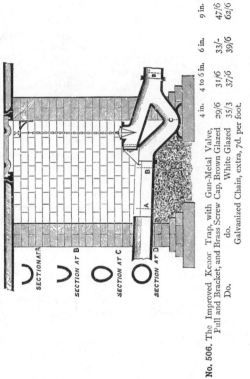

	4 in.	4 to 5 in.	6 in.	9 in.	12 in.
No. 506. The Improved Kenon Trap, with Gun-Metal Valve, Pull and Bracket, and Brass Screw Cap, Brown Glazed	29/6	31/6	33/-	47/6	92/6
Do. do. White Glazed	35/3	37/6	39/6	62/6	—
Galvanized Chain, extra, 7d. per foot.					

when I started loitering around public lavatories, in the name of research. I had discovered that yet another of Crapper's patents was Patent No. 3, 964, March 5, 1891 – T. Crapper, Automatically Flushing W.C.s. This was an ingeniously simple cistern which would flush urinals at regular intervals purely by siphonic action, without the complicated timing device which one would have expected. I wanted to see whether it was the same sort of system which was in use today but I envisaged the sort of trouble I, like Eric Coates, could be in if I turned to my neighbour in a public lavatory and said, 'Give me a leg up, will you, I want to have a look into this cistern.'

In the event, co-operative attendants at various conveniences helped me to confirm that the automatic flushing system today is to all intents and purposes the same as Crapper's 1891 version.

As evidenced by the legend 'T. CRAPPER, CHELSEA' still to be seen on display in many parts of the country, he was naturally involved in the world of urinals, with their splash guards, pet cocks and flush spreaders.

But although Crapper was to make numerous contributions to the betterment of the public convenience I do not feel that it is within the scope of this book to deal fully with that aspect. That field is in the more capable hands of Jonathan Routh. His *Loo Guide to London* might be described as the definitive book on the subject. When I last saw him he was preparing to fly off to New York and I asked him whether he had ever been there before.

'Never,' he said.

'Oh, well, you'll enjoy it. Seeing the Empire State and Fifth Avenue and all the other sights.'

'That's the trouble,' he said sadly. 'I'm being flown over to do a loo guide to New York and I'll be spending my time down there in the public lavatories.'

One must be frank and admit that not all of Crapper's developments met with immediate success.

d bathroom there was an accumulation of old
and fittings and other bits of plumbing which
o be carried downstairs. Crapper was always
nishing his men about 'the lazy man's load'
to carry in one trip what really required two
Frank never really got the message about this
n this occasion he set off down the steep stairs
is arms laden with the old cistern, lengths of
nd other lavatory bric-a-brac. He stepped on a
bit of carpet, his feet shot from under him
was only by sheer good fortune that when
nging and the clatter had subsided and some-
me to his aid on the landing he was found
ot much worse for his tumble.

n Crapper came to check on the job and was
out Astell's mishap he took a good look at
irs and pronounced the system of carpet
as downright dangerous. Astell had been
nough to come off unscathed, but a maid
a tray of breakfast things, or worse still,
carrying a baby... So, he gave the matter
ought and in due course came up with Patent
029, March 16, 1903 – T. Crapper, Stair
which on investigation at the Patent Office
to be not unlike a type still installed in
homes.

Tho. Crapper

Cheyne Walk and the river before the construction of the Embankment

What was called the Tro
have appeared, a dandy id
efficient flushing system
Schools, Workhouses, Fac
flushed the water down
('Quotations given for an
then into a trough und
underground stream, car
had not reckoned with
in the workhouses, who

Boys used to crumple
light to it and drop it
to have the occupants
howls of anguish as it
This made it necessary
Improved Trough Clos
each toilet had a wat
'all the objections hithe
are removed'.

Such a fertile mind d
his inventions were rig
tion. The matter of s
of his lighter stuff. T
his employees, Frank
very bright. But he
days, for instance, h
once to attempt the
valve on an outlet
at the mains, before
showering results. H
confined quarters of
wisely investigated a
and recoiling from
of his head on a hea
a week in bed w
spectacular was th
Crapper momentari
treads.
When Crapper

55

quate
pipes
had
admo
trying
trips.
and o
with
pipe a
loose
and it
the cla
one ca
to be n
Wher
told ab
the sta
treads
lucky e
carrying
a nanny
some th
No. 6,
Treads,
are seer
modern

56

Chapter 11
Ahead of His Time

The more research I did on Thos. Crapper the more I came to feel that he might well be termed the Barnes Wallis of the lavatory world. Just as Barnes Wallis was away ahead of his time in aircraft design, so too was Crapper in his field of endeavour.

The cantilever toilet, projecting from the wall with no means of support, is considered terribly modern. Referred to in the trade as 'wall hung', it is to be seen in the showrooms of the better sanitary ware suppliers and in the homes of wealthy trend-setters. Its advantages, of course, are that the cistern and all the pipes are out of sight and that the floor underneath can be cleaned unimpeded.

But the interesting thing is that there is nothing new at all about the cantilever toilet. Crapper had it on the market in 1888. His old catalogues show the detailed specifications for installing it, so that nothing is visible except the 'fireclay basin to build into wall' and the 'remote control chromium plated lever handle'. However, it was not in those days aimed at the carriage trade. It was developed for prisons and mental institutions.

When the flushing closet first came into general use it was not foreseen that prison and asylum inmates would find their own uses for all that metalwork in and around the cistern and yank from its fittings the ball valve, that long metal arm with the copper float on the end. It wasn't much fun for Matron, always going in dread of one of her charges creeping up behind her and giving her a bonk on the head with the big brass ball. It became a latter-day version of the bladder wielded by the King's Fool, but much more painful. Convicts went further than what the asylum inmates doubtless felt was just harmless fun. It was inviting trouble to have that chain dangling from the cistern, especially when it was a 'heavy duty brass chain' with a great 'china pull' on the end. Teddyboys with bicycle chains were nothing compared to prison guards being confronted with convicts swinging such lavatory chains. Also the prisoners could fashion some very lethal weapons from bits and pieces of metalwork plundered from the cisterns. So Crapper's No. 398 stored all that out of harm's way behind the wall and it is interesting that it was a safety measure not a piece of contemporary design that gave us the cantilever toilet which graces the more modern homes and well appointed new buildings such as New Zealand House in the Haymarket.

Visitors to Germany return raving about a type of toilet with no cistern which flushes straight from the mains with a really swingeing pressure of water. Trust the Germans, they say, to be advanced in this, as with so many other things mechanical. But they are not advanced at all. It is very old stuff. Crapper pioneered this more than half a century ago with his No. 3416 – The Patent Flusherette Valve.

The key to the whole thing is the Metropolitan Water Board. They don't approve. In the old Crapper catalogues featuring the Flusherette Valve there is a footnote: 'This Valve will not pass the Metropolitan Water Board's requirements.' It was for export only. The Water Board's tests revealed the danger of this type of flushing. With no cistern to interrupt the water supply, the water from the mains comes in direct contact with the 'soil' in the pan after the toilet has been used. We all know that electricity travels through water. (A friend of mine who took a great pride in his roses used to wire the plants with a small electric charge so that any dogs that lifted their legs on them got a dissuading surprise.) Likewise bacteria can travel through water and at a rare old speed – e.g. typhoid epidemics when water mains become contaminated by broken sewers. The Water Board has the public's well-being at heart in banning what would seem to be a really whizz-bang flush; the Germans are playing a lavatorial form of Russian Roulette by constantly allowing mains water to come into contact with 'soil' and keeping their fingers crossed that there won't be a bacterial jump.

One of Crapper's more unusual installations was at a house off Haverstock Hill. The Crapper Old Boy who told me about it said that it was not so much that the W.C. combination itself was unusual; it was the location. It was at the home of an artist named Lloyd, who wanted a toilet adjoining his studio. Part of a neighbouring room was partitioned off for this purpose and the window was increased in depth so that it was possible to relax and enjoy the view across the valley, which in those days was not built up to the extent it is now. Only the happy circumstances of it being on high ground with no possibility of being observed made this novel look-out a practical proposition.

It is rarely that one encounters what might be termed a loo with a view. The circumstances have to be special.

I understand that there are examples in a monastery on one of the Greek islands and at a resort in the Swiss Alps. In the latter case it is virtually a cantilever loo jutting out over a precipice and although the vista of snow covered mountains is breath-taking and disposal speedy and efficient many users find that vertigo can detract from what should otherwise be pure joy.

John Pudney writes of those with long memories of air travel agreeing 'that there had been nothing to equal the views obtained during transatlantic flights from the tail turrets of Liberators which, toward the end of the Second World War, were converted into smallest rooms. Here, surrounded by transparency, the sitter enjoyed a panorama from a vantage point, according to aeronautical savants, unequalled in aviation history.'

In *The Loos of Paris*, Jonathan Routh tells of a sad lost opportunity. 'It is on the *troisième étage* that we notice the most ridiculous feature concerning the Loos of the Eiffel Tower. However fantastic a feat of engineering it may have been to construct a loo 898 ft. above ground level, and even though the first person to use it may have been our own King Edward VII at its inauguration in 1889, it seems absolutely mad and ridiculous that this loo, which could have the finest view of any loo in Europe, has frosted glass in its square porthole of a window: and therefore no view at all.'

But no such opportunity was lost at Lord's cricket ground, practically in the heart of London.

Prior to the building of the new stand to replace the old tavern at Lord's, the most modern stand there was the Warner, built in 1963. It is a double-decker, for the use of members and friends, and one of the most popular parts of it is the long, glass-fronted bar and buffet between the lower and upper deck. One can stand there at the bar having a drink and watching the cricket – all day, if one has a mind to.

When the Warner Stand was being built Ronnie Aird, the then secretary of the MCC, did a tour of inspection one day and noticed that the three windows above the urinals of the men's toilets at each end of the bar were being fitted with frosted glass. It occurred to him that this was an unnecessary precaution. This was confirmed when he positioned various people at vantage points around the ground, to check lines of sight and, because of the overhang of the upper deck, none of them could see into the

A. Nursemaid's sink, discharging into B.
B. Slop-sink, with pottery basin and top and lead trap—a flushing cistern may with advantage be provided as shown for the w.c.
C. 2-inch lead branch anti-siphonage pipe from trap of B.
D. 2-inch lead main anti-siphonage pipe.
E. 3½- or 4-inch cast-iron soil pipe and drain-ventilating pipe.
F. 3-inch cast-iron rainwater pipe.
G. 2-, 2½-, or 3-inch lead soil pipe branch from B.
H. 1¼- or 1½-inch flush-pipe from 2-, 2½-, or 3-gallon cistern; the latter having overflow-pipe carried through the wall.
I. Pedestal wash-down w.c. with lead trap, and 3½- or 4-inch branch soil-pipe.
J. 2-inch lead anti-siphonage pipe from trap of w.c.
K. Enamelled fireclay-scullery-sink.
L. 1¼-inch lead anti-siphonage pipe carried through back wall.
M. 1¼- or 1½-inch lead waste pipe with P trap.
N. Trapped stoneware gully and channel receiving discharge from M and F.
O. Cast-iron bend with foot-rest at foot of soil-pipe.
P. Intercepting chamber receiving branch drains, ventilated by grating at top as an alternative, the iron cover may be air-tight, and a 4-inch pipe may be carried up the boundary wall and finished with a grating about a foot above the ground.
Q. Intercepting trap with air-tight stopper on cleansing arm.
R. Public sewer with junction-block for house-drain.

Notes: If the distance from the building to the intercepting chamber exceeds 3 or 4 yards, a small brick chamber with air-tight cover ought to be built under the area to receive the drains from N and O.

The drains under the area wall must be protected from injury by suitable lintels or must be of cast iron. Concrete foundations, equal in thickness to the external diameter of the drain and in width to three times this diameter, must be laid under each drain, and some local authorities require the pipes to be either embedded in concrete to half their depth or entirely surrounded with concrete.

0 1 2 3 4 5 6 7 8 9 10 feet

SANITARY FITTINGS AND DRAINS

toilets. So he got the workmen to replace the frosted panes with clear glass.

The view of the field from the urinals is, if anything, better than from the bar and thus it is that, through Ronnie Aird's initiative, it is possible to drink all day at the Warner bar, take time out for necessary interruptions, and yet not miss a single ball that's bowled.

I can think of no other sports ground in Britain which has been so thoughtful in such matters.

DOWN-RIGHT COCK.

No. 34

Chapter 12
Seating Accommodation

Crapper's patent No. 11,604, January 13, 1863. was for Self-Rising Closet Seats. Most older people can remember at one time or another encountering these tip-up jobs, activated by counterweights at the back. Crapper no doubt felt that he had hold of a cracking good idea to offset the problem which crops up everywhere except in nunneries and at the YWCA. But although much was made of the Self-Riser in his Catalogues when it came out and it was still featured in the 1930s, it became a discontinued line in the post-war era.

I asked a Crapper Old Boy why its appeal had dwindled and he said: 'All right for factories and schools, and some of the railways took it up. But no self-respecting family would feel the necessity of it. And old ladies objected to the likelihood of a smack on the bottom due to bad timing.'

Another Crapper idea which, if unveiled at next year's Ideal Home Exhibition, would create something of a sensation was his Patent No. 3,964, March 5, 1891. Can you visualise a toilet with no pull chain, no lever or switch of any sort with which to flush it, yet flush it does? I can see someone who is really trendy proudly showing off such an acquisition to a guest. 'But how does it work?' his friend would ask. 'You shall see,' our pace-setter would say, withdrawing discreetly.

The patent, of course, was Crapper's Seat Action Automatic Flush. It was interesting to look up the

Specifications at the Patent Office and see the 1891 drawing there, with the wording underneath: 'Front elevation of a pedestal water closet having my invention applied thereto.' Followed by all sorts of detail about how *a* slots into *b* and when the seat is depressed it retracts *c* etc. It was very ingenious and one might well wonder why it is not, like other Crapper developments, now standard equipment in our loos. But the simple answer is expense. As pointed out elsewhere, although people will splurge when it comes to such items for the home as a super refrigerator or washing machine, they are tight fisted in the bathroom. Why go to the added expense of this contrivance merely to activate the cistern?

As well as his tip-up seat and this one with the built-in flushing device, Crapper turned his attention to the shape of the seat itself and his 'pear-shaped, lip fronted closet seat' comes in for high praise in Ernest G. Blake's *Art and Craft of the Plumber*. Of 'best quality polished mahogany' and therefore worth safeguarding, this male oriented design was of course the precursor of what is now the gap fronted seat.

Crapper, the perfectionist, would have been horrified at what has happened to the toilet seat. The modern seat is a prime example of the fact that what is new and different is not necessarily better than what it replaces.

In Crapper's pre-plastic world, toilet seats were, of course, made of wood. They gave evidence of a nice bit of class distinction. Seats for the family were of glistening polished walnut or mahogany; those for the servants were of untreated white pine, known in the trade as 'scrubbables' for the very good reason that it was the job of the lowliest of the servants regularly to go to work on them with the scrubbing brush. But even if the family had a polished mahogany and the servants got a raw deal, those wooden seats were efficient. They were hinged to a baseboard securely anchored to the wall on iron brackets and 'secure' was the operative word. But modern designers of lavatory seats felt that they had to change all that. They felt bound to take something simple and efficient and unnecessarily complicate it.

Two holes were incorporated into the rear of the toilet bowl itself, to take two bolts with what are called 'moulded pillars', into which fits a rod on which were threaded the two projecting 'eyes' added to the back of the seat – the whole involved four-piece unit in plastic.

Since plastic bolted to a glazed pottery never has been or ever will be as secure as a metal bolt in wood, the modern plastic toilet seat in no time starts to drift off its moorings. At worst the whole seat can slide off and give you a nasty spill, and even shortly after it has been newly fitted you can have a disconcerting sense of insecurity at that time when, especially using the facilities in the home of a stranger, you are at your most vulnerable.

Whether there is a link between slithering toilet seats and the feeling of insecurity which causes the youth of today to turn to demo's, drugs and defiance, is something to which the psychiatrists will doubtless soon give their attention. In Victorian times everyone was more firmly based and no one can deny what solid citizens the Victorians were.

My younger son, who is at prep school at Melrose in Scotland, informs me that when they go to the *duffs* there is a cry of 'I bags the wooden one' and a mad scramble for occupancy of the toilet with the wooden seat, rather than having to use one of the four plastic-seated toilets. 'It's warm,' is his simple explanation. Since the *duffs* are by way of being what he describes as 'a social centre' at the school (they can enjoy illegal fish-and-chips and comics there), the boy who bags the wooden one is regarded as having an advantage over the others. So, to even things up, they have prized the lock off the door, which means that although he has the best seat in the house there is the drawback that his enjoyment of it can be interrupted.

Dr Johnson summed the whole thing up when he said: 'No, sir, there is nothing so good as the plain board.'

But Sir Winston Churchill did not agree with him. At his house at Hyde Park Gate although there were seats on the toilets for guests, his personal loo had no seat. When the plumber who did the work at the house said surely he would like to have one fitted, Churchill replied: 'I have no need of such things.'

No. 852

Chapter 13
Edith Crapper: A Memoir

Miss Edith Crapper, who is now nearing 80 years of age, is the last to bear the name Crapper of those connected with the firm. Her father George, a nephew of old Thomas, had spent all his working life in the business.

When I went to visit her she lived alone in the old family home in Bolingbroke Grove, Wandsworth Common. As she took me through into the living-room it was like going back over the years and stepping into a Victorian parlour. Pictures in the manner of Landseer hung from a picture rail. There were cane chairs and lace curtains. A piano. The corner cupboards were crowded with a wide variety of nick-nacks and there were statuettes under glass domes. The two armchairs and the sofa had dust covers over them. Apparently I didn't rate these covers being taken off and I wondered who did.

Tall, well spoken and alert, with a sense of humour always bubbling up to the surface. She is an artist and after she had served me tea from a silver teapot in which it had been correctly left to infuse before pouring she showed me some of her work. Exquisite miniatures, illuminated lettering, illustrations of religious themes, artwork for traditional Christmas cards sold by the galleries. Proudly she laid out a selection of personal Christmas cards. For many years she had done these for Lord and Lady Woolton.

Then she said, rather sadly. 'It isn't easy, you know.'

'What isn't?' I asked.

'Trying to make the name Crapper famous as an artist rather than a cistern.'

Ever since she had been a young girl when Queen Victoria's reign was drawing to a close Miss Crapper had been keen on art. One of Crapper's carmen, as they used to call the men who made the deliveries on horse-drawn open drays, was a man named Lush and he knew of her youthful passion for painting. Whenever his rounds took him to the Wandsworth area he used to buy a penny paintbox and make a special call to give it to her. With his dray laden with lavatory pans, cisterns and flush pipes he used to pull up at the house and she was always out there to meet him. As well as her excitement at getting a new painting set, it was also a chance to see Bonnie – Lush's horse – whom she loved. But Bonnie was a bit nippy, so when she fed him lumps of sugar she had to hold them out on a tennis racket.

Children are always interested in what their father does and this applied to young Edith. Her mother used to take her by horse bus to Harrod's for tea. I don't know how it was actually worded, but when tea was over her mother used to say something like, 'What would you like to do now, dear? Shall we go and watch Daddy testing toilets?'

At all events, Miss Crapper remembers going with her mother around to the Marlboro' Works, which was no great distance from Harrod's. There she could watch her father and other employees standing by as old Thomas pulled away at the chain of a *Deluge* or an *Alerto* to see whether some new modification to his cistern was practical. A Victorian outing.

But her youth was not all fun. One thing that clouded it was her relationship with the Wharam girls. The original Wharam was in charge of the business side of Crapper's. His sphere was the office. He had little knowledge of or interest in the plumbing side. He never got his hands dirty. Edith's

68

Mrs Langtry, the Jersey Lily

father, like old Thomas, was dedicated to the perfecting of the cisterns and allied sanitary products which bore the Crapper imprint. Young Edith used to feel terribly hurt when she was at a tea party, sitting on a sofa with the Wharam girls, they would edge away from her, not wishing to be associated with a Crapper, which after all was the source of their wealth.

When the art teacher at the Francis Holland School for Girls in London went away on extended leave, Miss Crapper was asked to fill in for her.

'When I was introduced to the girls they got the giggles,' she told me. 'I found out why later, when the headmistress was showing me around the school. In the girls' changing room was a row of four toilets, all with "Thos. Crapper, Chelsea" on them.'

I asked Miss Crapper if she was ever embarrassed by the name.

'Good heavens, no,' she said. 'Only the other day I was in Westminster Abbey, and there beneath my feet was the name Crapper.'

'In Poets' Corner?' I asked.

'No,' she said. 'Actually it was on a manhole cover. He did the drains for the Abbey, you know.'

Chapter 14
By Any Other Name

Robson Barrett joined Crapper's in 1904 as an office boy and when he retired as managing director he was in his sixtieth year with the firm. 'I have always wondered,' he said to me, 'why toilets in the old days were always given a name.' If he doesn't know the reason, I can't think of anyone else who would. But the fact remains that each type of W.C. in Victorian times had its 'private badge', as it was called in the trade – names like the *Optimus, Orion, Cedric, Planetas* and so on.

There were so many of these (I have accumulated 81 without even trying) that just as there are train spotters there are Toilet Spotters – people who collect the names of loos encountered in the antiquated washrooms of old pubs and railway stations and other places that are happy hunting grounds for those interested in Victoriana.

But the romance has gone out of toilets. Today they are just bleakly stamped with the name of the maker, and one orders particular models by number. This prosaic modern approach is presumably supposed to be more efficient, but in fact it isn't. Figures are much more prone to be misread than names. You are likely to order a 7526 and find yourself getting a 7528 by mistake. But in the old days when you ordered an *Improved Infante* there was no likelihood at all of your being lumbered with an *Original Burlington*.

If one studies the art of toilet naming as carried

70

Crapper and his employees in front of the Marlboro' Works

on in Victorian times one sees that many of the names fall into definite categories and the five main groups are these:

(1) THE 'HARD SELL' NAMES, such as the *Deluge, Cascade, Tornado* and *Niagara.* These were obviously aimed at the slice of the market which demanded that a toilet should be seen and heard to do a good flushing job.

(2) THE SPEED AND EFFICIENCY GROUP, like the *Rapido,* the *Alerto* and the *Subito.* I think *Alerto* is a wonderful name for a toilet. It conjures up a picture of it standing there poised to leap into instant action at the very touch of the chain.

(3) THE COMPOUND NAMES. These are made up from taking part of the manufacturer's name and part of the location of his works, as for example the *Twycliffe* (made by Twyfords of Cliffe Vale) and the *Sharcote* (Sharpe Brothers of Swadlingcote).

(4) CLASSICAL MYTHOLOGY. This gave rise to obvious ones like the *Pluvius* and the *Aquarius.* More subtle were the *Aeneas* (he got his association with water through being reared by a Sea Nymph) and the *Nereus* (he was a sea diety, son of Oceanus).

But by far the largest single category are—

(5) CRAPPER'S STREET NAMES. His toilets include the following:

Marlboro	*Onslow*
Walton	*Lennox*
Ovington	*Manresa*
Cadogan	*Culford*
Sloane	

These are all Chelsea streets, avenues and squares immediately adjacent to or within a stone's throw of the site of his original works in Marlborough Road, which has now been incorporated into Draycott Avenue. It is possible to walk in a continuous line from that site along *Walton* Street, down *Ovington* Street, though *Lennox* Gardens and across *Cadogan* Square to *Sloane* Street. It might well have been a lunchtime stroll which he took regularly.

If Crapper, a lover of the Royal borough, had had literary learnings he might have been prompted to write the type of book which one sees in the local library – *The Highways and Byways of Old Chelsea* (privately printed). Instead, like the character in Anatole France's *Juggler of Our Lady*, it appears that he decided to pay tribute to Chelsea in his own way. In so doing I shouldn't think he got any thanks from people who had been at pains to establish residence in such places as Cadogan Square, Lennox Gardens and Sloane Street because they represented 'a good address', only to find that they wound up as lavatory names.

The Victorian era in which Crapper spent most of his life was of course notorious for being prim. Women's breasts were 'the upper part of the body'. Even legs, which would seem to be a straightforward enough statement of fact, were often as not 'the nether extremities'. We have seen earlier that the Victorians did such a circumnavigation around 'toilet paper' that it became 'curl papers'. When it came to mention of the toilet itself and the necessity to go there the art of euphemism reached its apogee.

But the interesting thing is that today, in what we are invited to believe is the permissive society, we are no more direct in this regard than the Victorians were.

I was amused by a letter in the *Daily Express* from a woman who derided people for using euphemisms such as 'powder my nose' and 'spend a penny'. 'Why don't they come right out with it,' she wrote, 'and say "Where is the lavatory?"' The use of euphemisms is so ingrained in us that the dear soul was asking us to be blunt by using what in itself is a euphemism, a lavatory of course being a washbasin.

Death and drunkenness are two things we are at great pains to avoid referring to directly. People *kick the bucket, pass on, go the way of all flesh* or, in the classic Hollywood cliché, wind up by someone asking of them 'Is he ... ?' People get *tight, plastered, stoned,* have *one over the eight.* With little difficulty one can think of about thirty euphemisms for death and for drunkenness. But this is as nothing compared to the circuitous lengths people go to to avoid mention of the water closet.

I've jotted down sixty words and phrases for it and this is just scratching the surface. We go to the *loo, biffy, chamber of commerce, holy of holies, cloakroom, shot-tower, smallest room* etc., etc. We announce we want to *be excused, pick a daisy, see the geography of the house, wash our hands, turn our bicycle around, see a man about a dog* etc., etc.

Dominic, young son of Clement Freud, found himself on the front page of the *Daily Mail* when he was setting off with his father on the Trans-Atlantic Air Race and announced, 'I want to go somewhere.' Not meaning New York, of course. And every second being precious he was left behind to go somewhere.

It is quite impossible to list all the W.C. euphemisms in general use for the simple reason that people are so co-operative in helping others to skirt the subject that they will accept *any* word or phrase for it, provided your intent is made clear with a questing look.

There are of course a wide variety of regional and foreign variants. On the continent the telephone is very popular. It was common for members of the Resistance during the second World War to say: '*Je vais téléphoner à Hitler.*' In Denmark women ask the whereabouts of the *dametelefonen* – the ladies' phone. In *Fractured French,* that delightful book which translated *carte blanche* as 'For heaven's sake, somebody take Blanche home,' the expression *tant pis tant mieux* is 'My aunt feels better now that she has made that phone call.'

France, although described in a recent issue of *Plumbing Equipment News* as 'always laggard in sanitary sophistication', had been quick to acknowledge Britain's leadership by incorporating *le water-*

closet into their language immediately this country gave the W.C. to the world, modifying it in polite society to *les waters*. But the French as a general rule are much more open about this matter than we are. John Pudney tells of a friend who prided himself on his French going up to a *gendarme* and asking: '*Où peux-je aller pisser?*' To which the policeman, with a gesture which took in all the 207,076 square miles of that fair country, replied: '*Mais, monsieur, vous avez toute la France.*'

In Italy, hotels which find *W.C.* too blatant use the term *numero cento*, which can at first be rather bewildering to foreign visitors who notice that the room numbers jump from, say, No. 24 to No. 100. The whole aspect of the labelling of men's and women's toilets in hotels, bars and similar establishments is a study in itself. In New York there is a bar which carries subtlety to the ultimate by having one breed of dog on the men's and another on the women's. When one's need is pressing it can be quite aggravating to have to wait until it dawns that one is a pointer and the other a setter. The magazine *Private Eye* gleefully reported that it had come across an item in a Gloucester newspaper to the effect that the town council of Cheltenham Spa had voted to replace the words MEN and WOMEN on their lavatories with LADIES and GENTLEMEN in order 'to attract a better class of person'.

New Zealand is the most far-off country in the world, for the good reason that if you try to go any further than New Zealand you start coming back again. Yet in that isolated spot, long before the white man came, the native Maoris had followed the same *smallest room* train of thought as Europeans when evolving a euphemism. The Maori word for it is *whare-iti* – 'the small house'. When the need arises to be more specific, they say *whare-noho* – 'the sit-house'. New Zealanders frequently make up 'fractured' Maori words, such as naming a noisy speed boat *Wai-ki-kupa-rau*, and some hotels out there

74

delight in bewildering visitors to the country by labelling the toilet *Here-it-are*.

The natives of South Africa, similarly to New Zealand's Maoris, gave rise to their own version of the smallest house, *piccanniny kiaha*, being generally used for a toilet in that country. It is abbreviated to *P.K.*, which created a problem for the export department of the Wrigley Company of America, one of whose biggest selling lines had long been 'P.K.' chewing gum.

Australian males speak of going to the *dike*. A family term is the *proverbial*, which is an abbreviation of, stated politely, *the proverbial brick outhouse*. Australia being, like America, a land of great open spaces, the Chic Sale type of privy has persisted longer than in smaller countries such as England with far more built up areas served by sewers. An Englishman I know made a trip recently into the backblocks of Australia and in the course of it he happened upon a pub where he was surprised to see a trough half-way up the front of the bar and running the whole length of it. He did not need to ask the reason for this; observation soon told him. Since pubs of that type are all male affairs there was no likelihood of women being affronted but it struck him that it was all very functional, to say the least. He enquired of the publican how it came to be put there and learned that 'I rigged it up, sport, because of the money I was losing by customers wasting good drinking time by going out all the time to the *proverbial*.'

Since the United States is such a large country which has so many variants of speech, one should not generalise, but it is true to say that the term *loo*, so favoured in England, is not commonly heard over there. The word used by the same type of people in the States is *john*. It has a fine old history, dating back to at least 1735, when among the regulations issued at Harvard University was: 'No Freshman shall go into the Fellows' John.' For many years the

British firm Armitage, one of the biggest in sanitary-ware, enjoyed a great advantage over their rivals in export to the States because they used to trade under the name of their original owner, the Rev. Edward Johns. It appealed to Americans that they could literally buy Johns imported from England. *Rest room* gets a great deal of use in America and visitors find it odd that even at a gathering in some-one's home a guest will ask, 'Can you tell me where the rest room is?' But my favourite of the euphem-isms encountered in the U.S., and Canada, is *comfort station*. It has a ring about it which carries one back to the Elizabethans' *place of easement*.

And in America, of course, they have the *crapper*.

In Britain it is odd that despite all the names that were given to toilets and their manufacturers' names being prominently displayed, rarely have the British borrowed these for W.C. euphemisms. I did once know a girl who used to speak of going to have a *shark*, which on enquiry I found derived from the fact that the ancient loo in the flat shared with two other girls was a 'Shark' (No. 1316 in Crapper's old catalogues). But this is a rare and very localised example.

It is intriguing how the term *crapper* came into general use in America.

Soldiers on overseas duty are great ones for pick-ing up local words and bringing them back home. An apropos example is calling a lavatory a *karzie*, from the Hindustani word *khazi* which British troops who had served in India brought back with them.

American servicemen stationed in England during World War II took home the expression 'You've had it'. Their fathers in World War I took home *crapper*.

The firm old Thomas established did a lot of work in military hospitals and barracks in the first World War. The troops could not help but be made aware day after day that the sanitary arrangements were by T. Crapper, Chelsea. If one casts one's mind back to the 1914–18 period one realises that at that time in the hinterland of America they were not very advanced in what the *Plumbing Equipment News* calls 'sanitary sophistication'. American doughboys from the hills of Kentucky or an Idaho farm, where the privy was all they knew about, would be mighty impressed by Thos. Crapper's slick contraption. And they would be quick to pay verbal tribute to his inventiveness.

Thus it was that returning American troops spread the word. Going to the *crapper* got a great deal of use during the 1920s, so that eventually the *Dictionary of American Slang* was able to report that the term 'had come into common usage in America by 1930'.

The only reason one can think of why British troops didn't follow suit is that the Crapper insignia was old stuff to them. It had been around for many a long year in this country which, after all, was the pioneer of this whole water closet business. It was no novelty to them.

The word *crap* was a fine old English word. The book *Slang & Its Analogues*, published in 1891, gave it a variety of unrelated meanings – 'to harvest', 'the gallows', 'printers' type in disarray' and 'to ease oneself'. But nowhere in that book or any other English dictionary I have searched through can I find the word *crapper*. The Americans alone have given the man his due.

And from that word they have derived others. To quote from the *Dictionary of Americanisms:*

crappy – inferior, ugly, cheap, merchandise of inferior workmanship, inferior entertainment, *crappy* workmanship.

crap – nonsense, cant, lies, exaggeration, mendacity, bull. 1939: 'Pally, I never heard so much crap in such a short time in my life.' John O'Hara, *Pal Joey.* 1949: 'I'm not interested in stories about the past or any crap of that kind.' Arthur Miller, *Death of a Salesman.*

It is *crap* in this connotation which has drifted

across the Atlantic from America to England in recent years. One hears it from time to time in this country, as in 'You're talking a lot of crap.' But the English are hesitant about using it in polite society, associating it in their minds with the good old Anglo-Saxon word.

Presumably *crap*, meaning nonsense, will also become acceptable here – a strangely round-about way of his fellow-countrymen at long last giving credit to Thos. Crapper, the W.C. pioneer.

Chapter 15
Paper Work

The secretaries at Humpherson's, the sanitary equipment firm in London, have antique wooden envelope holders on their desks. Of beautiful workmanship in polished mahogany, they are admired by visitors. But what visitors don't know and new secretaries don't know is that they are not really envelope holders at all. They are Crapper toilet paper boxes (1s. 9d. the pair, in the 1872 catalogue).

The Humpherson who founded the firm almost a hundred years ago sent his two sons to Crapper's to learn the trade (their framed indentures still hang in Humpherson's offices today) and they always carried a full range of Crapper products. The present chairman of the company, Geoffrey Pidgeon, has a fine sensitivity about the old days of the loo and when he came across the old Crapper paper boxes he saw to it that they were not only preserved but put to useful purpose.

At Twyford's in Stoke they have similarly preserved old toilet paper holders, theirs being of highly glazed pottery in ornate designs of rich blue and gilt.

Of course in former days, up to the latter part of Queen Victoria's reign, the toilet roll as we know it today was something yet to come. It was always paper squares, either improvised from printing matter or made specially for the need, and although one might have thought that it was so, there was not a sharply defined distinction between the upper classes using purchased toilet paper and the lower classes relying on the printed word. In 1747 in his classic *Letters to his Son,* Lord Chesterfield wrote: 'I knew a gentleman who was so good a manager of his time that he would not even lose that small portion of it which the call of nature obliged him to pass in the necessary-house; but gradually went through all the Latin poets, in those moments. He bought, for example, a common edition of Horace, of which he tore off gradually a couple of pages, carried them with him to that necessary place, read them first, and then sent them down as a sacrifice to Cloacina; thus was so much time fairly gained; and I recommend you to follow his example. It is better than only doing what you cannot help doing at those moments and it will make any book which you shall read in that manner, very present to your mind.'

But apart from those who regarded it as a seat of learning, the customary thing was either squares in an attractive wooden or porcelain container, or accommodated on the more homely nail in the wall. In the latter category the Drayton Paper Works, of Wandsworth, stole a march on the rivals by bringing out cut pieces 'Ready Stringed'.

However, no matter what form it took, it was never referred to as 'toilet paper'. It was one of the numerous 'unmentionables' of Victorian times. At that time it happened that women were curling their hair by rolling it around folded-up bits of paper and this provided them with a useful euphemism when faced with the embarrassment of having to buy a fresh supply of toilet paper. The shopping housewife would ask the storekeeper for some 'curl papers'.

It is not certain to whom credit can rightly be assigned for the invention of the ingenious perforated toilet roll, although the date of its advent is known. In 1880 the British Patent Perforated Paper Company came into being and at around the same time two Philadelphia brothers called Scott, a name to become famous in the tissue world, turned from the marketing of wrapping paper to meeting the demand for this new fangled style of toilet paper.

It was not long before the Americans had graduated from the type of paper first used in rolls, which was to all intents the newsprint used in newspaper production. In 1907 crêped paper, the forerunner of the modern soft tissue, came into use in the States. But the British stayed faithful to 'Bronco', the name which the British Patent Perforated Paper Company had adopted as less unwieldly than their full title. Even in 1932, when two brothers named Rosenfelder fled from the growing menace of Hitler to set up a crêpe paper mill in Walthamstow, the softer crêpe version of toilet paper did not make any great inroads into the British preference for Bronco and similar brands known in the trade as 'hard tissue'.

Toilet paper is not a big export item. 'Because of its bulk, you're exporting air,' as one member of the trade put it. But for some odd reason, Belgium is not only a fairly large importer of British toilet paper but there is a market there for Bronco, the only other country to see eye-to-eye with Britain in this regard. But it is not called Bronco in Belgium. It is marketed there as 'British No. 3'. The Bronco people can only conclude that Belgians buy it for snob appeal, like Harris Tweed, Jags and Yardleys.

But apart from this little Belgian splinter group the whole of the civilised world outside Britain is on soft tissue. It never ceases to amaze overseas visitors to this country when they encounter Bronco or other of the hard tissue brands to which the British are still faithful.

We all know where to go when we want Government publications, but a little known function of H.M. Stationery Office is the issue of toilet paper, for it is they who supply every Government building in the land with its T.P. – from the House of Lords down to the lowliest labour exchange. From their Norwich storehouses come 7,000,000 rolls per annum. Government employees use up 20,000 rolls a day. An interesting statistic.

It is all endorsed, as they say, 'Government Property' on each sheet. And it is all Bronco. The Government is among the last hold-outs for hard tissue. The younger, more progressive element in Government circles feel that in this regard they are out of touch with the feeling of the nation. There is pressure being brought to bear to go along with the modern trend. At committee meetings supervised by the Provisioning Officer there is lobbying to have the toilet paper contract revised. At this time of writing a decision was hourly expected from the Government to switch to soft tissue.

The BBC, as might be expected, are still holding out for the old-fashioned hard tissue, endorsed BBC PROPERTY. But if, through your diligent efforts or with the help of the dangerous crossing in front of Broadcasting House which is nicknamed 'Promotion Corner', you graduate up to the Director-General's floor, you are not only entitled to use the executives' loo but you find yourself on soft tissue (pink).

Chapter 16
Personally Speaking

Crapper was of medium height, genial and kindly. When his youngest nephew Harry, the good-looking one, 'went off the rails', as they used to say, he gave him a job and despite the head-shaking of the family in general kept reinstating him over a period of five years because he knew Harry and his wife were in poor circumstances and he wanted to help them.

He invariably wore a billycock, that late Victorian piece of headgear that was a cross between a bowler and a top hat. Robson Barrett, last managing director of Crapper's, who had known the Old Man, or rather had been politely in awe of him, when he had started as an office boy in 1904, says that a noticeable thing about him was that he was always neat about the neck, his shirt spotlessly clean and immaculately pressed, the ends of his bow tie tucked under the collar in the style young men of the 1950s were to affect for evening wear. His grey beard was of the type that was to become known as the George V beard. He always wore a frock coat, as befitted his station in life, and that essential part of the uniform of the Victorian gentleman, a watch chain across the bulging waistcoat. On one end a gold hunter, the sort of watch with which you could make a nice little performance of checking the time as you flicked open the cover; on the other end a sovereign case or a seal. And to complete the picture, a walking cane, of course.

In 1867 he married Maria Green, whose Christian name was pronounced not in the West Side Story way but as in Victorian times – ma-*righ*-ah. She was a cousin with whom he had grown up in Thorne. He had always had a boyhood liking for her, they had kept closely in touch when he came to the big city, and when he felt he was sufficiently well established to undertake marriage he asked her to come up to London to be his wife. She arrived with her trousseau in a little wooden chest, most of it hand-sewn and costing £4 10s. in all.

She bore him one child, a son christened Thomas to carry on the name, who died in infancy. If one goes to Somerset House to look up the records and find out at what age the new Thomas Crapper had died one sees beside his name the entry so frequently encountered in those days before the Welfare State: 'Age – 0 years.'

Mrs Edna Brown, a grand-niece, remembers her as a very kind person. She recalls happy Christmas days as a little girl at the Crappers' home in Thornsett Road in south-east London on the edge of Kent. 'The turkey midday dinner was followed by a high tea in the evening when the ham, served hot with the turkey, appeared again cold plus brawn and a whole Stilton cheese.' As a special treat she was allowed to watch the Crystal Palace fireworks from one of the bedrooms upstairs. With no children of her own, Maria loved youngsters and Edna Brown remembers the pleasure she got from lifting her little grand-niece up to talk for the first time into the new miracle of the age, the telephone. At such times as Easter and Whitsun young Edna would be 'taken for a whole-day drive by hired carriage into the country, with all of us getting out and walking on the hills.' In 1902, on her tenth birthday, she received a letter from Great-aunt Maria in which she wrote, 'Now you're into double figures!' It was the last letter she got from her. Maria died that year in an old-fashioned sounding way. 'One day when she was on a visit to old Thomas at his works she had a fit of coughing to which she succumbed.'

There were two servants at the Thornsett Road house, the cook and Ellen, who identified herself as the undermaid by letting the strings of her cap hang loose, as was the habit of those days. She was frail. Soon she was to become so ill that not only was there the need for a doctor but also a specialist and a lengthy term in hospital. At that time before the Health Service, old Thomas paid for it all. In her eighties, and living in Scotland now, she said in a letter: 'I owe my life to Mr Crapper, a very likeable Gentleman who always thought of the working class. He was a big engineer and, as you probably know, the inventor of the water Lav, which was a pull-up plug before that.'

Old Thomas loved gardening and as well as the elaborate greenhouses at the bottom of his garden at Thornsett Road he had an extra garden plot in the then sparsely built neighbouring street, Wheathill Road. Mrs Brown remembers that on Sunday visits as a girl with other youngsters of the family Great-Uncle Thomas used to take them around to show them the garden. A humorous soul, he would tease them by prolonging their eagerness to get back to the tea and cakes to be provided by Great-Aunt Maria as he identified the various blooms for them by their Latin names and then, when at last they felt they were heading back to the house, he would pause again, to prod the manure pile with his walking cane and say, 'Ah, the lovely smell of the country.'

The Crappers are spread far and wide. There is a grand-niece of old Thomas in Canada by the name of Minnie Finch who has 18 grandchildren and has done her best to carry on the strain if not the name of Crapper.

Brought up using the quill pen, Crapper remained faithful to it for most of his life. In the 1850s a Birmingham firm started the manufacture of a steel pen 'which will supersede the quill as the recorder

of our wants, our business and our affections' and
began turning them out by the gross from their
'manufactory'. But just as in modern times many
people resisted the incursion of the ballpoint and for
precisely the same reason, Crapper was among the
many Victorians who were loath to see an end to
the glorious thicks and thins of writing by quill pen.
But although the results were a joy to the aesthete
one always had to have a knife handy since the con-
founded things were in constant need of sharpening
and we don't stop to realise that it was because of
this earlier use that it came to be known as a *pen-
knife*.

Although he had come to London as a boy there
were still traces of his Yorkshire origins in his speech.
Sugar, for instance, he called 'shugger'.

He was a Mason.

Robert M. Wharam, whom he had brought down
from Yorkshire in 1867 to look after the business
side of his firm, was a quite different kettle of fish.
He was a strict Wesleyan, non-drinking, non-
smoking, what has been described as the 'if it's fun
legislate against it' type. He was the supreme pessi-
mist. Every day of the year he carried an umbrella.
There was a pub which backed on to the yard of
the Marlboro' Works and the workmen used to have
pints of beer smuggled to them through the wrought
iron grillwork that divided the two properties. 'Old
Wharam died in ignorance of this,' commented a
former employee. He was outstandingly unlikeable.
When the offices and showroom were moved to the
King's Road he used to drop over to the Works
from time to time to keep tabs on how things were
going. On one occasion one of the office staff had
slipped over to have a pint with the lads and was
almost caught red handed. In the rush for cover he
climbed over a brick wall and surprised a good lady
in her back yard. 'What's wrong?' she asked. 'Is
the old bastard about?'

However, Wharam was a shrewd businessman – 'cute' in the old-fashioned English meaning of the word, certainly not the modern Americanisation of it. Crapper was the ideas man, the visionary, the experimenter. He found paper work tedious, consuming of time he felt he could put to much better purpose designing, say, a new type of drain ventilator. So he was content to have Wharam look after the accounts and the other office routine. Also he felt a certain loyalty to him, for they had been friends as youngsters up in Yorkshire.

When he was getting on in years old Thomas did not go into the office each day, leaving it very much to his nephew George to run the practical side and Wharam the management, and himself dropping in only about once a week.

However, before actually retiring he felt that he would like to see what had now been built up into a thriving business housed in a more dignified building than the old Marlboro' Works. After a good deal of searching he managed to find a good property available in the King's Road, at No. 120.

So, in 1907 the move was made and with the business he had built up from scratch now prominently on display in an important London thoroughfare Thomas felt that the time had come for him to put his affairs in order and lead the quiet life in his comfortable house and attractive garden down on the border of Kent.

His regret was that he had no son to leave the firm to. There were nephews, of course, through whom he could perpetuate the name of Crapper, Sanitary Engineer, Chelsea. But Charles and Tom were doing well enough for themselves as builders and decorators in another part of London. There were other nephews strewn around Yorkshire and out in the colonies, but it was at too late a stage now to bring them in. Of course, there was still Harry, the good-looking one, but he had proved himself incapable of even holding down the sinecure that had been provided for him and by now was so far off the rails that hope had to be abandoned of ever getting him back on the permanent way. And there was nephew George Crapper, who was on the payroll, in charge of the manufacturing side. A good man but not quite with the business acumen necessary to run the whole firm.

So old Thomas came to the decision, or was prodded to it, that the thing to do would be to pass the business on to Wharam. He had, too late, second thoughts about this one day when he over-heard one of his staff say to a companion as they observed Wharam arriving for work: 'Here comes Pontius Pilate.'

Crapper never set foot in the building again.

Pushes.

No. 1141.

Chapter 17

The End of the Road

When Crapper's wife died in 1902 two maiden nieces, Emma and Sarah, came to live with him at Anerley in south-east London. The house still stands at No. 12 Thornsett Road, a street of not unattractive two-storey Victorian houses. It is occupied now by a retired dentist, B. S. Sherwood, and his wife and when I dropped in they told me that they had heard of the Misses Crapper but they were long since dead when they took the house over in 1945. But there was ample evidence of the Crapper association; all the bathroom and washroom installations were his products. Naturally he would have the best, including a bath which didn't have a plug but instead a pull-up waste pipe, which was *the* thing in those days and according to the catalogues a £3 15s. extra. The bath was encased in panelled mahogany. The toilet seat cover was cork topped, the most expensive of the entire range of that period. I was disappointed that I was not able to see these. Mrs Sherwood told me that a few years previously they had modernised everything.

But she was able to be helpful in another respect. She told me that in the early part of this century the house next door had been occupied by W. B. Yeats. I found it interesting that Crapper had been a next-door neighbour of the illustrious Irish poet, although nowhere in his autobiographical writings does Yeats make any mention of this fact.

Crapper's life fitted neatly into two reigns. He was

born when Queen Victoria came to the throne, in 1837, and died when her son Edward VII died, in 1910.

He was very much confined to his bed towards the end and one day he was disturbed by Emma having difficulty with the W.C. chain in the bathroom adjoining his room. The repeated ineffectual pulling of the chain, a familiar sound in those days and still to be heard, was something especially familiar to old Thomas.

One can imagine the mental turmoil he went through, lying there listening to something he knew he could put to rights in a trice. It brings to mind the composer who was similarly bedridden and was driven to distraction by someone in the room above finishing his piano practice on an unresolved chord; the composer could stand the anguish no longer and had to stagger from his bed, upstairs to resolve the chord.

It was the same with old Thomas. He could not bear the sound of that inefficient chain. Allowing a decent period to elapse until his niece had abandoned her efforts and returned to her household duties, he got from his bed and crept into the bathroom. His practised fingers soon found the flaw in the mechanism, the toilet was flushed and he returned to his bed.

It is tempting to say that he then died, a dedicated craftsman to the end. But although it was not his final gesture it did in fact happen near enough to the end for it to be imprinted on the minds of those who survive him.

A short walk from Thornsett Road is Elmers End Cemetery, where Crapper is buried. It adjoins the main railway line to the Kent coast, which is the reason twelve bombs landed in its 40 acres during the last war, the German air force being eager to sever this line of communication with Dover and other Channel ports.

At the lodge at the entrance they looked up

Thomas Crapper for me in an old ledger in which the entries had been made with a quill pen – Plot 4165 V4 Row 1. The plot had cost 10 guineas, I noticed.

The pathways and flower beds around the lodge and along by the chapel and the new graves were immaculately kept but when I turned off to walk among the older graves the paths were of crumbling asphalt, many of the headstones were askew, weeds were the main item of foliage, small shrubs were growing up out of the drains. I don't know much about cemeteries; perhaps this is the usual thing. But then I came upon a grave that stood out from its untidy, depressing surroundings by virtue of being beautifully cared for, the big white marble cross and the whole of the plot kept as neat and clean as if it had been newly there. Crapper's grave? No. A plaque proclaimed: 'W. G. Grace. Doctor and Cricketer. 1848–1915.'

I had not known that the great 'W.G.', a Bristol man, had been buried at Elmers End. Later I was to learn that he lived in nearby Beckenham in his declining years and there is a pub adjoining the cemetery called the 'Dr W. G. Grace'. Its saloon bar contains all sorts of mementoes of the Old Man – the bat he used in the first ever Test against Australia in 1884, a souvenir silk handkerchief bearing a portrait of the 'Champion Cricketer of the World', a letter he wrote to a Mr Page asking him how many games he would like to play for Gloucestershire in the 1886 season. I asked the manager how long the pub had been called the 'W. G. Grace'.

'Ever since it was built,' he said.

'When was that?' I asked.

'Two years ago.'

At the cemetery, when I had paid my respects to the memory of cricket's first great crowd-pleaser I walked on in search of Crapper's grave. I did not need to go far. It was only a matter of a few yards and there was his grave, shared with his wife. It is what is technically known as a 'gabled ledger', i.e. lying flat. On the one side was, 'In loving memory of Maria, the beloved wife of Thomas Crapper, who died July 1st, 1902, aged 65 years' and on the other, 'In loving memory of Thomas Crapper, who died January 17th, 1910, aged 73.'

It was plain and simple. On Grace's grave, what the man is remembered for is symbolised by three cricket stumps, a bat and a ball. The same sort of thing could have been done for Crapper, but I'm sure it was felt that it would have looked undignified.

I went for a stroll along the winding paths of the cemetery, thinking about old Thos., how he had made a fair contribution, from the time he had come to London as an 11-year-old, to rise to the top as master-craftsman and inventor and to go to his grave as Royal Plumber.

I passed the Grace memorial again on the way out and as I came away I felt that it was not in-appropriate that two great Englishmen who had given pleasure to so many should be buried in neighbouring graves.

The Crapper building at 120 King's Road, a con-version of a three-storey Georgian mansion, was familiar to all the old timers of Chelsea. Since it was the headquarters of a firm more than 100 years old, they were sorry to see it vacated in 1966 when John Bolding and Co, the take-over company, transferred the Crapper interests to one of their own buildings and it ceased to exist as a separate firm.

No. 120 King's Road stood empty for a couple of years, the main entrance boarded up, the display windows painted over. But at that time, when I went down to have a look at it I saw that someone, clearly a sentimentalist, had left the name – THOMAS CRAPPER in bold lettering – on the façade. But of the passers-by, who bothered to look up, either literally or metaphorically, to that name

W. G. Grace at the wicket, 1906

writ large above the door? Not a thought was given to it by the Chelsea young men with their long flowing hair and their three-quarter length jackets of purple, maroon and pale fudge. The girls tripping by in their mini skirts paid it no heed.

However, Joseph Losey, director of that excellent film *The Servant* had regard for it. The building is the focal point of the opening sequence of the picture.

Today it is occupied by, among other things, a shop called SKIN. This terse name applies to the fact that they are in business selling trendy suede and other leather gear. More than one Crapper Old Boy has told me that he shuts his eyes when he passes the shop.

Adjoining the building is a coffee-bar restaurant called Guys-n-Dolls which displays a notice: 'Special breakfast – 9 a.m. to 12.' It is perhaps just as well that Thomas Crapper did not live to see the day when a restaurant would cater to people who have breakfast anywhere between 9 and 12 o'clock. Toiling away to get the cistern right for future generations, he would have done a day's work by 12 o'clock.

I walked up from Chelsea to Westminster Abbey to see if, as grand niece Edith had mentioned, his name was indeed to be seen there. I kept a lookout for names on manhole covers I came to on the pavement, in the hope of seeing some original Crappers. There were plenty of HAYWARDS LTD LONDON S.E.1. Others were by S W FARMER LONDON S.E.13. and there was one which, if there are such people as manhole cover spotters, must be a rare collector's item – A SMELLIE LONDON W.C.

But none of them, sad to say, was the product of our man. When I got to the Abbey, however, and went across Dean's yard into the body of the building I was not let down.

One walks along the cloisters and there beneath one's feet are the names on the tombstones ... Edward Tufnel 1719 ... William Postard 1201 ...

The blind scholar Andrew Fisher 1614, Author of Defence of the Liturgy...William Sheffield Musician and Composer 1748–1829. And then, even if it *is* only on a manhole cover, Thos. Crapper, Sanitary Engineer Chelsea.

His name lives on.

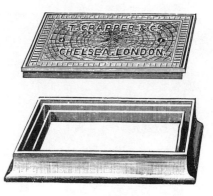

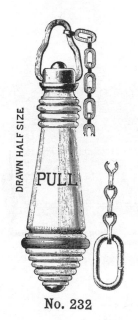

DRAWN HALF SIZE

PULL

No. 232

Illustrations by courtesy of
Edith Crapper, Radio Times Hulton
Library, Chelsea Public Library,
Twyford's and from Crapper Catalogues